POISONOUS PLANTS
AND ANIMALS
OF FLORIDA AND THE
CARIBBEAN

DAVID W. NELLIS

PINEAPPLE PRESS, INC.

SARASOTA, FLORIDA

Inquiries should be addressed to:
Pineapple Press, Inc.
P.O. Box 3899
Sarasota, Florida 34230

Library of Congress Cataloging in Publication Data

Nellis, David W.
            Poisonous plants and animals of Florida and the Caribbean / by David W. Nellis.
— 1st ed.
            P.   Cm.
Includes bibliographical references and index.
ISBN 1-56164-111-1 (hardcover : alk. paper). — ISBN 1-56164-113-8 (pbk. : alk. Paper)
            1. Plant toxins—Florida. 2. Plant toxins—Caribbean Area. 3. Animal toxins—
Florida. 4. Animals toxins—Caribbean Area. 5. Poisonous plants—Florida 6. Poisonous plants—
Caribbean Area. 7. Poisonous animals—Florida 8. Poisonous animals—Caribbean Area.   I. Title
RA1250.N45 1996
615.9'5'09759—dc20   96-21971
                                        CIP

First Edition
10 9 8 7 6 5 4 3 2 1

Design by Carol Tornatore
Printed in Hong Kong by Sing Cheong Printing Co. Ltd.

DISCLAIMER
This publication is intended to acquaint the reader with the appearance and characteristics of poisonous plants and animals that can be encountered in Florida and the Caribbean. Neither publisher nor author claim that the list of organisms described herein is exhaustive of harmful organisms that may be encountered in the geographic region covered. This book should not be relied upon as a source of medical advice. The reader should seek immediate medical attention if poisoning or suspected poisoning occurs.

Table of Contents

Table of Contents

Table of Contents

ANIMALS

Introduction

Why do we have poisonous plants?

Poisonous plants can be found in every nation on the world. Toward the polar limits of vigorous vegetation growth there are few actively toxic plants. As one moves toward the equator the variety of plants in the landscape greatly increases and the competition between plants becomes more intense. Animals that consume plants also become more abundant and vary from leaf miners that live in and consume only small portions of single leaves to elephants which consume many trees in a day. Even slight advantages conferred by toxins in reducing attention by herbivores can make the difference between survival and extinction. The tropical rain forest with many species actively competing for survival contains a cornucopia of plants and animals that use toxins to give them an edge on the competition. The distribution of toxic organisms is also a result of past evolutionary and geologic history. The notoriously toxic Euphorbia family is predominantly tropical while the Solanaceae are distributed over temperate and tropical habitats. In Florida and the Caribbean we have only modest numbers of toxic plants compared to other jurisdictions. The percentage of snakes that are venomous is much lower in Florida and the Caribbean than in Australia and other areas. The most dangerous and potentially lethal jellyfish, cone shells, octopuses, scorpions, and centipedes are not found in this part of the world.

The African connection

Many of the plants in the American tropics are also found in tropical West Africa. While some, such as the ackee, were intentionally introduced by ecological entrepreneurs, others came as incidental passengers on the many slave-trading ships or as seeds incidental to other cargo. A more ancient and widespread source of this transatlantic distribution is continental drift. In the geologic past when South America and Africa were part of the same land mass, plants were uniformly distributed by habitat. As the rift which was to become the Atlantic Ocean gradually widened, many of the plants continued to exist on both sides of the ditch. As the continents moved further apart and world climate changed, the West (moist) African distribution of many species has been restricted to the Gulf of Guinea and the Congo River drainage basin, while the American distribution has spread from southern Mexico and the West Indies to Peru and Brazil.

Folk medicine

Modern medicinal treatments have developed very gradually over several thousand years. Some of the earliest Egyptian records made reference to medicinal plants. Theophrastus, a student of Aristotle who wrote on plants, based much of his work on data collected from a group of citizens called *rhizomatomi* who studied and prescribed plants for their medicinal properties. Pedanos Dioscorides, an ancient Greek physician, wrote a book on medicinal plants which provided much of the basis for botany and medicine for fifteen centuries. During the Renaissance, books on herbal medicine blossomed and the field grew vigorously through the nineteenth century. Modern science as practiced in the twentieth century has greatly refined the art of using plants to treat medical problems by determining efficacy and standardizing doses of active ingredients. Many people still practice the personal collection of plant parts for medicinal use.

I am not recommending the folk medicinal use of the plants discussed in this book but rather am chronicling it as an interesting anthropological utilization of a natural product. Many of the older generation now living in the West Indies had little access to modern medicine until after World War II. While many plants in folk medicine have no known valid scientific effect other than as a placebo, others are known to have potent influences on the body and the course of disease. Both effects were fully utilized and in some cases are still in use in more remote areas and in lesser-developed countries.

Modern medicine requires that commercially produced drugs be safe and effective and of a consistent potency. In folk medicinal potions, the margin between safe, effective dose and harmful dose may be too narrow for acceptance by modern regulatory agencies, and the plant constituent responsible for the effects may be completely unknown. Most of the plants used in folk medicine have not been scientifically tested for their reputed pharmacological activities and fewer still have had their active components identified. When natural products are collected and utilized for medicinal purposes, the potency of their active ingredient may vary greatly. The quantity and quality of an active component may vary considerably depending on environmental conditions. A plant growing in the shade in the rich, moist soil of a valley bottom may have only 10 percent of the alkaloid concentration of the same species growing in the sun in the poor, dry hillside soil nearby. Different races or cultivars of plants may show widely varying toxic propensities.

One use of toxic plants that is probably generally effective is for elimination of intestinal worms. The toxin incapacitates the parasites and loosens their hold on the intestinal walls at the same time that strong intestinal contractions are induced as the body responds to the plant toxins.

It is almost certain that many folk medicinal uses of plants have a pharmaceutical basis. As traditional herbal medicine declines with "development" of many primitive people and with the passing of the older generation in modern societies, these folk medicinal clues to medically useful plants are being lost to mankind.

Information sources

This book has been prepared from published literature, interviews, and personal observations. The fiscal realities of publishing prevent me from citing and listing the two thousand journal articles and 250 books consulted in preparing this compendium. The somewhat irregular coverage of technical and chemical information about the listed plants is due to insufficiencies in our knowledge. Even some of the most common weeds and ornamentals have not been thoroughly investigated. The people who willingly shared their experiences and traditions are kept anonymous to protect their privacy. As a tropical field biologist for twenty-five years, I have had the opportunity to personally feel the adverse effects of many of the organisms listed here, and most of the others have been confirmed by first-hand testimony.

Treatments

Few plant toxins have specific antidotes. The best-known treatment for many plant toxins is to remove the remaining plant material from the body as rapidly as possible. In the case of external irritation, this is usually best done by washing the skin with soap and water or rinsing the eyes with clear water. Many plant toxins induce the body to try to eliminate them by vomiting. The vomiting reflex is highly effective in eliminating undesirable compounds from the stomach. For plant toxins which do not induce vomiting, it can be reliably produced by a dose of two tablespoons of syrup of ipecac (one tablespoon for children under twelve). It is usually best not to induce vomiting when someone is unconscious or having convulsions, or has consumed corrosive or highly irritating compounds. A physician can perform gastric lavage to remove corrosive compounds from the stomach without further damage to the esophagus or danger of inhalation pneumonia.

Purging with cathartic substances eliminates the plant from the intestinal tract, reducing the time available for absorption of toxins. Two or three ounces of activated charcoal mixed in a glass of water and drunk adsorb many toxins, making them less available for uptake by the body. Cathartics and activated charcoal, used together, often complement each other's action in preventing further uptake of poisonous substances in the gastrointestinal tract. Cathartics should not be used to treat poisoning by corrosive substances and should be avoided if a person's electrolyte balance is disturbed or if he or she has a kidney malfunction. Using castor oil or mineral oil as a cathartic may increase the absorption of fat-soluble poisons.

Cyanide poisoning from eating plants which contain cyanogenic compounds can produce a life-threatening situation in a very short time. The symptoms include dizziness, headache, vomiting, and difficulty breathing.

The blood pressure will be high and the pulse rate initially slow, then increase and become rapid. Convulsions, coma, and respiratory arrest may occur. The patient should be given pure oxygen for breathing and rushed to a hospital for treatment (an injection of 300 mg sodium nitrate followed by 12.5 g sodium thiosulfate). Further dosage will depend on hemoglobin levels.

In the case of plants which irritate the membranes of the mouth and throat with their calcium oxalate needles (raphides), relief is often obtained by consuming acidic household compounds such as lime juice or vinegar. The pain is often reduced by sucking on ice.

Mechanical injury
Many plants in our area have spines, thorns, prickles, and stiff, pointed hairs that cause mechanical injury. Cactuses often have clusters of barbed hairs called glochids at the bases of the spines. The small glochids and hairs produce local patches of inflamed papules (circumscribed, red, elevated areas of the skin) which may to lead to granuloma (granular tumor) formation in the skin. When the body eventually expels the foreign particles, a discolored spot often remains. Glochids and plant hairs may be removed with peel-off facial gels. Applied and allowed to dry, these gels remove large numbers of small, offensive particles with minimal fuss. Alternatively, a cloth saturated with a water-soluble white glue can be spread over the affected surface and allowed to dry. Removal of the cloth will remove the embedded particles. In every case, removal of the intruding plant part is easier if it is done promptly before swelling takes place.

Thorns and spines embedded near joints may produce a chronic or acute inflammation of the joint. When deeply embedded in or near bone, the puncture may heal but the delayed response may be similar to that of a bone tumor, with either erosion of or addition to the bone visible on X-rays.

The spines of many cactuses have barbs that allow only forward movement, so they may migrate slowly through the tissues of the body and emerge or lodge at some distance from their entry point. They should be extracted with a plucking motion to reduce prolonged discomfort. Other spines and thorns should be removed by grasping firmly with forceps and pulling them gently parallel to their axis.

A boon to mankind
Plants have been used medicinally by humans since the earliest hunter-gatherer societies. For three thousand years, organized societies have had individuals who specialized in using plants to treat the medical complaints of clients. Generations of trial-and-error experimentation based on guess and superstition have produced an extensive herbal pharmacopoeia. While ignorance and chicanery have resulted in the application of worthless remedies, much comfort—and some cures—have resulted from the use of herbal

medicine. Organic food enthusiasts have created a new market for many old herbal remedies which are "free of chemicals" but actually owe their effects to the great variety of chemicals produced by plants themselves.

One man's panacea may be another man's poison. The common willow tree provided generations of herbalists with a reliable analgesic which is marketed today as aspirin, a very safe and beneficial industrial copy of the original drug. Yet every year youngsters are killed by overdoses of aspirin.

Today, plants are a source of many alkaloids and commercial pharmaceuticals which can be extracted at less cost than they can be synthesized in the laboratory. It has been stated in the technical literature that only ten percent of the pharmaceutical compounds obtained from plants today can be commercially synthesized at competitive costs. The drugs extracted from plants have frequently served as models or source material for the chemical synthesis of other similar drugs. Synthetic drugs are usually the result of laboratory modification of a compound first found to be efficacious from plant sources. As an example, the pharmaceutically useful alkaloids from *Rauwolfia* served as the model for the synthesis of such common tranquilizers as Chlorpromazine, Milton, and Equanil.

Is it poisonous?

It is not possible to determine the edibility of plants by watching which ones are eaten by animals. The plant toxins which harm mankind do not necessarily adversely affect other mammals and birds. Ruminant animals such as goats and cattle have a veritable chemical factory inside them in which food is predigested by a myriad of microorganisms. These microorganisms frequently reduce toxic compounds to innocuous substances before they are further digested and absorbed by the mammal. Other animals have a sufficiently different physiology that they experience no adverse effects from eating plants poisonous to man. Over seventy-five species of birds intentionally eat poison ivy berries, and many species of wildlife eat mushrooms which would yield grave results if consumed by humans.

The color of a fruit is no guide to its edibility, as there are examples of both toxic and edible fruit of almost every shade. The taste and smell of unknown fruits provides no guide to their safety. Some of the most harmful fruits are sweet and delicious; their evil symptoms do not appear for hours.

The individual response to a particular toxin varies greatly. The same berry that produces serious symptoms in a twenty-five-pound child may not produce any effect on a two-hundred-pound man. A young, vigorous person may be able to detoxify or withstand the effects of a toxin while an older person in poor health may not have sufficiently efficient liver and kidneys to eliminate a toxin, or his body may be less able to withstand the adverse influence of a toxic substance.

The best policy when meeting a tempting fruit tree for the first time is to treat it as poisonous until confirmed otherwise. This concept instilled strongly in young children could prevent thousands of poisonings a year.

Plant chemical defenses

As animal life evolved on earth, by necessity it existed by consuming plants. As plants continued to evolve, they developed certain secondary metabolic compounds that proved distasteful to animals. The plants producing these toxic or repugnant compounds were able to reproduce more successfully and became the ancestors of the plants we see today. Plant metabolism has followed such a cornucopia of biochemical pathways that even relatively closely related plants may have a wide divergence in the chemicals present in their tissues. The toxicity of specific compounds to different animal groups varies widely, with many chemicals seemingly developed with herbivorous insects as the target and only incidentally or accidentally toxic to vertebrates.

Some plants produce latex as a first line of defense in healing wounds to their bark. Latex has the added advantage of containing various noxious or toxic substances which usually discourage further injury to the plant by feeding herbivorous animals. Other plants produce secondary chemicals or byproducts which do not appear to have any significant role in the metabolic function of the plant, but may have adverse effects on mammals consuming or having contact with the plant. Other plants seem to have evolved specific devious and ingenious methods for inducing discomfort in the animals which disturb them. Following is a brief discussion of some of the groups of chemicals produced by plants which are toxic to mammals.

Alkaloids are the most common of plant toxins and are present in detectable amounts in about 10 percent of all plants. Usually nitrogen-containing cyclic compounds, they are basic in pH and usually have a bitter taste. They characteristically exert their pharmaceutical influence in small doses. Some examples are the pharmaceutically important tropane alkaloids atropine and scopolamine found in jimsonweed, the pyridine alkaloids characterized by nicotine found in tobacco, and the indole alkaloids found in Carolina jessamine and psilocybin-containing mushrooms. The purine alkaloids are responsible for the physiological effects of coffee, tea, and cocoa. Many beneficial pharmaceuticals are plant alkaloids. The pyrrolizidine alkaloids found in several herbal teas may produce a sometimes fatal liver disease.

Phenanthrene compounds include steroidal sapogenins, saponins, and various steroid alkaloids which are cardioactive or cardiotoxic. Examples include solanine in the deadly nightshade and the many cardioactive compounds found in luckynut.

Glycosides are two-part molecules which yield a sugar and a pharmaceutically active compound when broken down by hydrolysis. Cyanogenic glycosides such as the dhurrin of sorghum break down to yield hydrocyanic acid (cyanide) and sugar. Cardioactive glycosides such as oleandrin exert their influence directly on the heart muscle, but also may produce other symptoms related to decreased heart function. Saponin glycosides break down to yield a soap-like compound which alters the permeability of cell membranes, producing many systemic disturbances. They are rapidly absorbed by the gills of fish and are frequently used as fish poisons. Anthraquinone glycosides are typically cathartic.

Oxalic acid and compounds containing it called oxalates occur as relatively soluble forms (with sodium and potassium)which produce a sour taste, as in rhubarb stalks, or systemic toxicity in animals when present in greater amounts. The insoluble forms (with calcium) often form sharp, needle-like crystals called raphides which physically damage the tender surfaces of the mouth and throat. The mouth and throat irritation that occurs after eating the underripe fruit of pineapple and *Monstera* are due to calcium oxalate raphides.

Phenols are hydroxyl groups connected to aromatic rings which frequently are irritating and produce allergies. The phenols present in *Toxicodendron* and other *Anacardiaceae* produce severe allergic contact dermatitis, while the phenol gossypol from cottonseed oil is a promising male contraceptive.

Tannins are complex polyphenol mixtures with molecular weights between five hundred and three thousand which are typically concentrated in the bark, roots, or leaves. Tannins may render many plant parts bitter and unpalatable or produce medical problems such as diarrhea or kidney or liver damage due to the alteration of animal proteins. There is substantial evidence that chronic consumption of high amounts of tannins is associated with the development of certain cancers. Yet much of the world's human population regularly consumes the tannins present in beverages such as tea and red wine with seeming impunity.

Terpenes, resins, and volatile oils are produced by many plants and are held in special ducts or cells. Terpenes have a chemical formula which is a multiple of C_5H_8, while resins and oils are an assemblage of complex compounds that usually include terpenes in their structure. These compounds aid the plant in repelling pathogens and herbivorous animals by producing various toxic and irritant responses. They usually are soluble in organic solvents. The active compounds in the *Euphorbiaceae* are in this group and frequently are cocarcinogenic in that they induce cancers in cells which have been exposed previously to very low levels of carcinogens. Many of our spices and flavoring compounds are in this group of chemicals, and have added considerable enjoyment to many lives when used in moderation.

Proteins and peptides are composed of amino acids and are constituents of all living cells. Toxic proteins composed of hundreds or thousands of amino acids are called toxalbumins or phytotoxins. These include the most toxic natural compounds known, such as abrin and ricin from *Abrus* and *Ricinus* respectively. Polypeptides are smaller compounds, usually containing less than a dozen amino acids. They are represented by the hypoglycin of ackee and the phenylethylamine and tyramine of mistletoe berries. Some of the most lethal mushroom toxins are polypeptides.

Names

Plant classification continues to change as new data are gathered. Added to this are varied taxonomic opinions put forth by taxonomists specializing in a particular group of plants. To reduce this situation to some fixed common denominator, I have used *A Synonymized Checklist of the Vascular Flora of the United States, Canada and Greenland* by Jon T. Kartesz as the universal authority for the Latin names of plants used in this book.

For zoological names I have used the Latin and common names which seem to be most prevalent in peer-reviewed journals.

Acknowledgments

Dr. Richard Schultes, the noted scholar of pharmaceutically active plants, provided advice and encouragement early in the preparation of this book and reviewed the completed manuscript. Dr. Dan H. Nicolson and Dr. R. A. Defilipps of the National Museum of Natural History, Smithsonian Institution, have provided many useful suggestions, particularly on the derivations of the Latin binomials. Marielle Brandon has reviewed and contributed to the account of *Cassiopeia*. Toni Thomas and Eleanor Gibney helped resolve some confusions on identification and helped locate specimens for photography. Dr. Robert Breene invited me into his house to meet his many pet spiders, including the *Loxosceles* used as an illustration. Lloyd Davis reviewed the manuscript and provided much lore on fire ants and a remarkable eye for capturing snakes to serve as models for some of the photos. Dr. José H. Leal of the Bailey-Matthews Shell Museum on Sanibel Island loaned the cone shells illustrated. Sara Barnett and other librarians were of consistent help in finding obscure material. Finally, my thanks go to the patience and good nature of those who always provided a smile and acceptance of my interest and time spent doing the research and photography for this book.

PART I

Distribution, Appearance and Habitat

Dinoflagellata

Red Tide
Gymnodinium breve

p. 121
Dinoflagellata

Distribution This genus of about 200 species is found worldwide. Several other species are known to produce similar toxins. Red tides are caused by very high population densities of this organism. They have been recorded in the Gulf of Mexico, in the Southwest Atlantic north to Cape Hatteras and in the Caribbean.

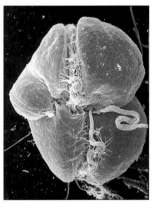

Florida Department of Natural Resources

Description The organism is a microscopic planktonic dinoflagellate about 30 microm-eters in breadth which floats freely in coastal waters. When circumstances are favorable, it becomes so abundant that it gives a reddish cast to the water; thus, the common name red tide. The calcium carbonate shell of the microorganism has two flagella, one in an axial groove and the other in a longitudinal groove.

Habitat The organism is not uncommon in open coastal waters, but sudden population "blooms" begin offshore. The maintenance and proliferation of blooms may be associated with increased nutrients associated with land runoff as a result of heavy rains. They are not usually found in significant numbers in salinities below 31 parts per thousand (seawater is about 35); thus, they do not normally penetrate into brackish estuaries except under conditions of drought.

Reproduction A complex life cycle includes the formation of inactive resting cysts on the sea floor which germinate to produce motile cells in the water column. These free-swimming cells reproduce by asexual division and compete with the other phytoplankton in the water.

Food Dinoflagellates produce their own food by photosynthesis.

Longevity A particular bloom may exist and drift for several months, but individual cells are more short-lived.

Ciguatera
Gambierdiscus toxicus

p. 122

Dinoflagellata

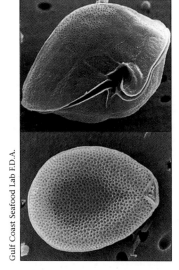

Gulf Coast Seafood Lab F.D.A.

Distribution This dinoflagellate has been found in all the tropical seas of the world. Its poleward extension seems to be limited to waters which support coral reefs. The local distribution is peculiarly patchy, with certain islands having abundant populations while nearby islands have none.

Description The clam-shaped calcium carbonate shell of this microorganism has a pitted surface and two flagella, one in an axial groove and the other in a longitudinal groove.

Habitat The dinoflagellate is found living among sessile filamentous algae which clings to rocks in shallow water.

Reproduction These mobile algae usually reproduce by division and thus produce clones. An increase in reported cases of ciguatera in the fall is correlated with warmer water, along with increased nutrients and decreased salinity associated with seasonally heavier rainfall. These dinoflagellates are often associated with early ecological colonization of shallow-water substrates disturbed by man or natural causes.

Food These microscopic organisms contain chlorophyll and produce their own food.

Longevity It is not known how long the individual dinoflagellates live under natural conditions (and it may not be a relevant question for an organism that reproduces by division), but populations seem to remain stable over periods of many years.

Fungi

Mushrooms

p. 125

Amanita species   Amanitaceae

Distribution   The mushrooms in this genus are common worldwide.

Appearance   The identification of mushrooms is a very specialized science. While several genera found in the wild have characteristics making identification for culinary purposes relatively easy, it is unfortunate that the most poisonous mushrooms are extremely similar in appearance to the most common edible forms. The genus *Amanita* has a distinctive veil or collar on the stem just below the cap. The cap shows considerable variation in

size, shape and color, but the gills are nearly always white. The common mushroom sold in grocery stores is an *Agaricus* grown under very carefully controlled circumstances and is completely safe to eat, but individual specimens of the deadliest mushrooms known to man may appear almost identical. Most knowledgeable mycologists refrain from eating field-gathered mushrooms of the standard mushroom shape characterized by the genus *Amanita*.

Because of the many tragic poisonings from misidentified wild mushrooms I recommend that you forage for exotic mushrooms at your vegetable market. The common white *Agaricus* is available from button to stuffing size. The Italian brown Crimini with a slightly scaly cap has a more pungent flavor, while the flat-capped Shiitake (which grows on old logs) has a rich, woodsy flavor. The Oyster mushroom with a fluted shape has a delicate velvety flavor. The slender creamy clusters of Enoki mushrooms with tiny caps have a slightly crunchy texture with a mild flavor. The large size, substantial texture and hearty flavor of the Portabella allow it to be used like meat in many recipes.

Vascular Plants

Sisal
Agave sisalana

p. 127
Agavaceae

Distribution The 300 species in the genus are found from the southern U.S. to South America. This species, a native of the Yucatan peninsula, is now cultivated and has escaped into many tropical countries. Several other species are also widely cultivated and still others are native to our area with only limited distributions. The medical and chemical considerations below generally apply to the genus.

Appearance This turgid succulent plant has a short, thick stem. Tough, fibrous roots radiate close to the surface to take advantage of the slightest rain in semiarid habitats.
Leaves The large, erect, dark-green leaves are stiff and spine-tipped with smooth edges growing to 1.5 m (5 ft) long. They form rosettes arising from near the base of the stem. *Agave picta* has dark-green leaves with a white or yellow band and straight spines on the leaf edges. *Agave americana*, with stiff, recurved spines on the leaf edges, has several cultivars; the one with light-gray leaves is the most common.
Flowers The yellowish-green flowers occur as lateral clumps on a central spike as tall as 6 m (20 ft).
Fruit The mature plant flowers once at about age 20. Reproduction is primarily by bulbils which germinate and develop on the seed stalk before falling to start many new plants near the parent. The rarely produced, 7 mm (0.3 in) seeds form in a thick, beaked capsule. This species also reproduces by suckers arising from the base of the parent plant before it blooms and dies.

Habitat This plant is able to thrive on rocky, dry sites on which few other plants can survive.

6

White atamasco lily
Zephyranthes atamasco

p. 128
Amaryllidaceae

Distribution The 50 species in the genus are found in the warm parts of the Western Hemisphere. This species is a native of Mexico, Central America and Florida, but is widely planted elsewhere in the tropics. Several medicinally similar species differ primarily in the color of their flower: *Zephyranthes citrina* is found growing wild in Florida with a yellow flower, and *Zephyranthes rosea* with a pink flower.

Appearance A perennial, lily-like herb with a white-fleshed, brown-coated subterranean bulb about 3 cm (1 in) thick and a tuft of 4 or more linear, flat, bluish-green leaves.
Leaves The 4 to 6 green, linear, grass-like leaves grow to 40 cm (16 in) long by 6 mm (1/4 in) wide.
Flowers The single, erect, fragrant, bright-white, tulip-shaped flower is supported on a slender stalk as long as the leaves.
Fruit The 3-lobed capsule contains shiny, flat, black, D-shaped seeds.

Habitat It is most frequently found in periodically wet sites on recently cleared land and in grassy areas.

Cashew
Anacardium occidentale

p. 129
Anacardiaceae

Distribution The genus of 15 species of trees and shrubs is found in tropical America. The natural range of this species (perhaps spread from Brazil by Native Americans) includes the shores of the Caribbean from Florida and Mexico south to Peru and Brazil. It has been introduced and naturalized in Africa, India and many other tropical lands. Most of the commercial production of nuts is in Africa and Asia.

Appearance This plant has a dense but irregular crown and varies in size from a shrub to a small tree 10 m (33 ft) tall with a smooth, gray bark and a trunk 15 cm (6 in) in diameter. The tree grows very quickly and bears within 2 to 3 years of planting.

Leaves The leathery, alternate, elliptic leaves are a rich green with prominent yellow, sunken veins. They grow up to 15 cm long by 7 cm wide, but may be only half that size.

Flowers The fragrant, pink to light-purple flowers with 5 narrow petals are borne in clusters on the ends of branches.

Fruit The bean-shaped edible cashew nut is enclosed in a poisonous kidney-shaped shell growing from the end of a red-blushed or yellow, thin-skinned, fruit-like structure (called the apple, but actually a swollen stalk). The fruit is edible, slightly acid and astringent, but pleasantly aromatic with a juicy, yellow flesh. The fruit from trees which produce the yellow apples is generally more sweet than that of the red.

Habitat The tree does best in the lowlands on moist, sandy soils in full sun, but it can grow in a great variety of frost-free circumstances. It does not thrive in dry areas or when subjected to drought conditions.

Christmas-bush
Comocladia dodonaea

p. 131

Anacardiaceae

Distribution The genus of 20 species of shrubs or small trees is found in Central America and the West Indies. This plant is a native of the Windward and Leeward Islands and the Greater Antilles except for Cuba and Jamaica.

Appearance This open-canopied shrub, sometimes with a sprawling, vinelike habit, may grow to 6 m (20 ft) tall with an 8 cm (3 in) trunk. The clear sap congeals and turns black upon exposure to air.

Leaves The alternate, dark-green leaves are deeply toothed and often tinged with red in a manner reminiscent of Christmas holly. Each tooth of the leaf is tipped with a small spine capable of pricking the skin even when dry.

Flowers The 2 mm (.08 in), purple, male and female flowers are borne on a short stalk.

Fruit The orange-red, fleshy, 10 mm (0.4 in), elliptical fruits contain a single seed.

Habitat This tough, hardy shrub may be found growing in the sun on beach berms, and in dry and moist forest when exposed to full sun for at least part of the day. It is able to persist in poor soil on steep, dry, rocky hillsides.

Mango
p. 132

Mangifera indica
Anacardiaceae

Distribution The 40 species in the genus are native to Southeast Asia. The mango was originally a native of India and Southeast Asia, and has been widely introduced to all tropical lands in its 4,000 years of cultivation. Europeans brought it to Central and South America in the 17th century and it was widely dispersed in the New World by the 18th century.

Appearance A large tree to 25 m (85 ft) tall with a broad, dense, rounded crown and sturdy trunk to 1 m (3 ft) in diameter. It has a clear, resinous sap.

Leaves The large, alternate, leathery, lance-shaped leaves, 15 to 30 cm (6 to 12 in) in length, droop in clusters at the end of branches. The leaves of many varieties are a deep wine color when they first emerge.

Flowers The 6 mm (0.24 in), yellow, 5-petaled, fragrant flowers are borne as conspicuous, upright, branched clusters on twig tips.

Fruit The large [to 2 kg (5 lb)], aromatic, elliptic fruits with smooth, leathery skins of many colors, hanging on stalks outside the foliage, have a delicious, juicy, fibrous flesh enclosing a single, flat seed. Hundreds of cultivated varieties have been developed, with larger fruits of different shapes containing fewer fibers and various flavors of flesh.

Habitat The mango has been introduced and has escaped to occupy all but the most salty, dry or high-elevation sites. The tree does best in an area with strongly marked seasonality of rainfall with a dry season at fruiting.

Poisonwood
Metopium toxiferum

p. 134
Anacardiaceae

Distribution The several species in this genus are native to Florida, Mexico and the West Indies. This tree is found from South Florida through the Bahamas and Greater Antilles.

Appearance This small tree to 10 m (30 ft) tall with a trunk to 30 cm (12 in) has sturdy, spreading branches producing a broad, rounded crown. The brown outer bark is distinctive as it flakes off, revealing the yellowish-brown inner layer. The abundant, clear, sticky sap turns black and congeals on exposure to air.

Leaves The shiny, alternate, pinnate leaves usually have 5 leathery, oval leaflets with blades about 5 cm (2 in) long. The young leaves are silky purple.

Flowers The tiny, green, 5-parted flowers in slender clusters emerging at the leaf axils near the branch tips are quite inconspicuous, but very attractive to bees.

Fruit The single-seeded fruits turn from green to orange-brown at maturity, and have a thick, resinous flesh. They are oblong, 9 to 12 mm (3/8 to 1/2 in) in length and distributed in open, pendulous clusters on fruiting stems.

Habitat May be found growing on moist or dry sites, more frequently on soils underlain by limestone. It is a pioneer species which rapidly invades but is excluded by fire in pinelands.

Brazilian pepper
Schinus terebinthifolius

p. 135
Anacardiaceae

Distribution The genus of 30 species is native to South America. A native of Brazil and Argentina, this ornamental tree was introduced to the southern U.S. by the Department of Agriculture in 1898 and has now been spread throughout the tropics. The similar and weedy *Schinus molle* is also cultivated and has escaped in the Caribbean area.

Appearance A shrub or small tree to 15 m (50 ft) in height with a trunk to 1 m (3 ft) and a dense, oval crown.

Leaves The alternate, pinnate leaves, with a red midrib, are 10 cm (4 in) long and have 5 to 9 lance-shaped, glossy, aromatic leaflets with no petioles. The crushed foliage has an aromatic, turpentine-like odor due to the resin content.

Flowers The white, 3 mm (1/8 in), 5-parted flowers are in lateral clusters arising from the base of the leaves. The male and female flowers are on separate trees.

Fruit The brilliant-red, persistent, fleshy fruits, occurring in dense clusters, have a brown pulp enclosing a single, brown seed.

Habitat This tree has a considerable range of acceptable growing conditions and has become an aggressive, difficult-to-control, weedy pest in South Florida. It continues to be widely used as an ornamental tree and as a windbreak or hedge. The dispersal of the plant is greatly enhanced by viable seeds in bird droppings. Once established, the tree grows rapidly and is able to suppress plant competition by various allelopathic chemicals. The result is a habitat with reduced diversity which supports fewer native plant and animal species.

Poison ivy, oak, sumac
Toxicodendron radicans, toxicaruum, vernix

p. 136
Anacardiaceae

Distribution The genus of 15 species is found in the Americas and Asia. With considerable variation, this species is found from Canada to Mexico, the Bahamas and eastern Asia. The variable species *Toxicodendron toxicarium* (poison oak) and *Toxicodendron vernix* (poison sumac), also discussed below, are found on the east coast of the U.S.

Appearance A woody, climbing vine or shrub which clasps the substrate with small aerial rootlets. Often has red stems, seldom exceeding 2 cm (1 in) in diameter but capable of forming a trunk 15 cm (6 in) thick and a tree 6 m (20 ft) tall. *Toxicodendron toxicarium* is usually a woody shrub to 2 m (6 ft) tall. *Toxicodendron vernix* is a shrub or small tree to 10 m (33 ft) tall with an open, ascending, branching pattern.

Leaves The alternate, glossy leaves are composed of 3 leaflets, the terminal one with a petiole. The leaflets are quite variable in shape and often turn red in the winter. The poison oak derives its common name solely from the resemblance of the leaves to those of white oak. Poison sumac has alternate, pinnately compound leaves to 30 cm (12 in) long, sometimes with a red midrib and about 11 elliptic leaflets which become colorful in the fall.

Flowers The small, greenish flowers are loosely bunched on short, lateral stalks.

Fruit The grayish-white or pale-yellow, waxy, spherical berry is about 5 mm (0.2 in) in diameter with a single, striped seed.

Poison Ivy
Toxicodendron radicans
Anacardiaceae

Habitat *Toxicodendron radicans* is found in open woodlands and savannas, usually growing in open shade. *T. toxicarium* is usually found growing on coarse, well-drained sand in coastal scrub-oak forests. Although *T. vernix* may be found in moist, upland vegetation communities, it is most commonly encountered in swamps and relatively wet areas.

Poison oak
Toxicodendron toxicarium
Anacardiaceae

Poison sumac
Toxicodendron vernix
Anacardiaceae

Yellow allamanda
Allamanda cathartica

p. 138
Apocynaceae

Distribution The genus of 12 species of climbing shrubs and vines is found in tropical America. A native of Brazil, this species has been spread throughout the warm parts of the world by cultivation.

Appearance This climbing shrub or vine with sticky, white sap grows to 3 m (10 ft) tall or 15 m (50 feet) in length.
Leaves The dark-green, pin-nately veined leaves, tapering at the base and with an acute drip tip, are up to 15 cm (6 in) long and arranged in whorls at the nodes.
Flowers The bell-shaped, brilliant-yellow flowers are 5-lobed and up to 10 cm (4 in) wide. Horticulturists have developed several flower variations, includ-ing one with flowers 13 cm (5 in) across with a magnolia scent, and others with double sets of petals.
Fruit The rarely produced, 7 cm (0.28 in), bristly seedpod splits open to release the flat, brown, winged seeds.

Habitat The plant thrives in a broad variety of circumstances in full sun, but seldom escapes from cultivation.

Periwinkle
Catharanthus roseus

p. 139
Apocynaceae

Distribution The genus of 5 species is found in the tropics, particularly in Madagascar. This species, originally from Madagascar, is now cultivated and has been naturalized worldwide.

Appearance An erect, perennial, bushy herb to 80 cm (30 in) tall.
Leaves The elliptical, glossy, dark-green, opposite, 5 cm (2 in) long leaves have a lighter midrib and almost no petiole.
Flowers The showy discoid flowers, which may be rose, white or white with a deep red throat, are usually present throughout the year.
Fruit The small, inconspicuous pod, with many small, black seeds, splits open when ripe.

Habitat It grows along coastal sandy areas and is an early colonizer on disturbed sites and along roadsides. Groups of plants persist after cultivation by self-seeding until shaded out by a succession of taller plants.

Oleander
Nerium oleander

p. 140
Apocynaceae

Distribution The genus of 2 species is native to the Old World. A native of the Mediterranean, this shrub has been introduced and naturalized for several hundred years worldwide in the tropics and subtropics. In the U.S., it is found from South Carolina south to Florida and west to California. The literature reports of characteristics probably include the similar *Nerium indicum*.

Appearance   Usually a spreading shrub with many erect stems, but sometimes becoming a small tree to 6 m (20 ft) tall.
Leaves   The shiny, narrow, leathery, lanceolate, leaves (at least 15 cm (6 in) long) with many parallel lateral veins are in whorls on short stems.
Flowers   The white, pink or dark-red fragrant flowers occur in clusters on the branch tips beyond the leaves.
Fruit   The hairy, flattened seeds, each with a parachute tuft, are borne in narrow, curved, paired, 15 cm (6 in) pods. The pods are seldom formed in cultivated plants.

Habitat   With drought and salt tolerance and a deep, spreading root system, this shrub can grow almost anywhere and is widely cultivated in the tropics and subtropics. It is grown in large pots and tubs which can be moved inside in colder climates.

Bitter ash   p. 143
Rauvolfia nitida   Apocynaceae

Distribution   The genus of about 100 species is found throughout the world in the tropics. This species is a native of Central America and the West Indies, but it has now been introduced and naturalized in Australia and India.

Appearance   A shrub or tree to 12 m (40 ft) with a 40 cm (16 in) trunk and an open, rounded crown. _Rauvolfia tetraphylla_, a shrub to 5 m (16 ft) with similar chemistry, is often confused with this plant in the literature.
Leaves   The leathery, lance-shaped leaves are 8 to 15 cm (3 to 6 in) long and are whorled in groups of 4.
Flowers   The 6 mm,. white, tubular flowers occur as terminal or lateral clusters.
Fruit   The flattened, fleshy, spherical, red fruits 10 mm (0.4 in) in diameter with 1 or 2 rugose, hemispherical, brown seeds turn from green to red to black when ripe.

Habitat   This tree is found growing at low elevations in forests and scrub land.

Crape jasmine
Tabernaemontana divaricata

p. 144
Apocynaceae

Distribution The genus of 140 species of trees and shrubs is found worldwide in the tropics. This species is a native of India, but has been widely introduced in the warm parts of the world as an ornamental.

Appearance A symmetrical, spreading shrub to 3 m (10 ft) tall which produces latex from injured twigs.
Leaves The shiny, oval, lanceolate, 15 cm (6 in) leaves are arranged oppositely on the green twigs.
Flowers The 5 cm (2 in), white, ruffled, fragrant flowers are borne on stalks originating at the leaf bases. Cultivated forms often have double flowers.
Fruit The 5 cm (2 in), orange-red pods have curved beaks at the tip.

Habitat This popular decorative plant adapts to a variety of soil conditions in sun or open shade.

Luckynut
Thevetia peruviana

p. 145
Apocynaceae

Distribution The genus of 9 species is found in the American tropics. This small tree, a native of Central America, has been introduced and has escaped throughout the tropics.

Appearance A spreading shrub or small tree to 8 m (26 ft) tall with a 10 cm (4 in) trunk and a rounded, bushy crown. It exudes an abundant, sticky, white sap when wounded.

Leaves The slim, shiny, linear leaves, 7-15 cm (3-6 in) long, are pointed on both ends and have almost no petiole.

Flowers The waxy, yellow or orange, funnel-shaped, fragrant flowers are borne in a lateral or terminal cluster.

Fruit The pyramidal, yellow fruit, about 3 cm (1 in) wide, has a thin flesh covering the hard, brown stone which encloses 2 to 4 large seeds.

Habitat This tree reproduces readily by seed in a broad variety of habitats.

Dumb cane
Dieffenbachia seguine

p. 147
Araceae

Distribution The genus of 25 to 30 species is found in the American tropics. This species is native to the Greater Antilles and the tropical American mainland. Other similar or perhaps synonymous species commonly cultivated are *D. maculata, D. exotica, D. picta* and *D. amoena*. There are several hundred cultivated varieties. The toxic effects of all of these species seem to be similar.

Appearance A fleshy, usually erect herb with evenly spaced, horizontal leaf scars and thick, succulent stems to 2 m (6 ft) tall.

Leaves The elliptical leaves, with prominent midveins, have clasping, rounded petioles and blades which are mottled with ivory along the major veins and up to 50 cm (20 in) long. The petioles of *D. picta* are grooved or channeled.

Flowers The rarely formed flowers are minute and borne on an erect spike covered with a green to white sheath.

Fruit The fruits are bright-red, spherical berries which are rarely seen on potted plants.

Habitat This is a plant which likes the moist shade beneath the canopy of the tropical rainforest. It may be found growing wild in damp or fully saturated soil in the bottom of a forest ravine. It adapts well to the reduced light of being a potted house plant.

Pothos   p. 149
Epipremnum aureum   Araceae

Distribution The genus of about 8 species is found in Malaysia and nearby Pacific islands. This species is a native of the tropical Pacific. Its variegated cultivar has now been introduced worldwide in the tropics as an ornamental.

Appearance A rampant, vigorous, climbing vine to 30 m (100 ft) long and 10 cm (4 in) thick.
Leaves The heart-shaped leaves, often yellow-mottled and with clefts, may develop blades to 50 cm (20 in) in length.
Flowers The rarely seen flowers are densely packed on a short, vertical structure.
Fruit The fruit is borne in a vertical, cone-like structure.

Habitat This plant may be found climbing in trees or spread thickly over the ground in shady areas. It is widely cultivated indoors, where it maintains a more delicate demeanor.

Monstera   p. 150
Monstera deliciosa   Araceae

Distribution The genus of epiphytic climbers has 25 species in the tropical Americas. This species is a native of the tropical forests of Mexico, but is now widely cultivated outside in warm climates and in greenhouses.

Appearance A large, strong, climbing vine to 20 m (70 ft) which clings to its substrate with sturdy rootlets.

Leaves The huge, dark-green leaves have blades to 1 m (3 ft) in length with elliptic, pinnately arranged holes.
Flowers The small, densely crowded flowers cover a thick, vertical spike up to 30 cm (1 ft) tall flanked by a green spathe.
Fruit The closely packed fruits grow from a vertical, cone-like structure. The individual fruits on the cone ripen sequentially from base to tip.

Habitat The plant does best when growing in moist, fertile soil in open shade. It is widely grown in a pot and is used in interior decorating.

Aralia   p. 150
Polyscias guilfoylei   Araliaceae

Distribution The genus of 80 species of shrubs is found from tropical Asia to Polynesia. A native of the South Pacific islands, this plant is now cultivated in the warm Americas and as a potted plant in cooler regions. The taxonomy and species boundaries within this genus are somewhat uncertain.

Appearance An erect, compact, fast-growing shrub to 6 m (20 ft) tall with sparse, stiffly vertical branches.
Leaves The alternate, aromatic, compound, toothed leaves, to 40 cm (16 in) with 7 cm (5 in) leaflets, are often irregularly bordered with white.
Flowers The seldom-seen, small, white flowers occur on long branching stalks.

Fruit The inconspicuous fruits are seldom produced on cultivated plants.

Habitat It is grown as an ornamental and as a hedge.

Fishtail palm
Caryota mitis

p. 151
Arecaceae

Distribution The genus of 13 species is found in the Indo-Malayan region. This species is from tropical Asia. It and the similar *Caryota urens* are planted as ornamentals throughout the tropics.

Appearance This palm grows to 10 m (33 ft) tall, often in clumps arising from suckers. The similar wine palm, *Caryota urens,* grows to 80 ft and does not produce suckers.

Leaves The arching, bi-pinnately compound leaves up to 3 m (10 ft) long are divided into many segments, each resembling the tail of a fancy goldfish. This is the only genus of palms with bi-pinnate leaves.

Flowers The small, inconspicuous, pale-cream flowers are borne in long, branched inflorescences at the bases of the leaves.

Fruit The globular, 1.5 cm (0.6 in), orange to red fruits which turn black when mature are in multi-strand bunches resembling strings of beads near the leaf bases.

Habitat Although not tolerant of salt, it does well in dense shade on a variety of soil types.

Butterfly weed
Asclepias curassavica

p. 152
Asclepiadaceae

Distribution The genus of about 120 species is native primarily in the Americas. This New World tropical native has been widely introduced and naturalized throughout the tropics.

Appearance An erect, perennial, sparely branching herb to 1 m (3 ft) in height with a viscous, white sap.
Leaves The narrow, opposite, lanceolate leaves with conspicuous horizontal veins are 10 cm (4 in) in length.
Flowers The bright-red flowers, each with a yellow-orange central corona, are borne in umbrella-shaped clusters on a terminal stalk.
Fruit The 10 cm (4 in), tapered, cylindrical pods split open when ripe, releasing many flat, brown seeds, each with a parachute of silky fibers which carries it through the air with the slightest puff of wind.

Habitat As a rapidly colonizing, wind-spread weed, it is found in open pastures, disturbed sites and along roadsides in full sun or partial shade. It flowers in response to rain and is never found in the dense shade of forests.

Giant milkweed
Calotropis procera

p. 153
Asclepiadaceae

Distribution The six species in this genus are native to Africa and Asia. A native of dry tropical Asia, this large species has been introduced accidentally and intentionally throughout the tropics. Most of the characteristics below also apply to the similar *Calotropis gigantea*.

Appearance This plant is usually seen as a perennial herb, but may reach a height of 6 m (20 ft) with a trunk 25 cm (10 in) in diameter and thick, white, sticky sap. The smooth, light-green bark becomes light-gray on older parts of the plant.
Leaves The leathery, oval, opposite leaves are heart-shaped at the base and pointed at the tip, with prominent midrib and blades 25 cm (4 in) or more in length. The petioles are short and sturdy.
Flowers The succulent, white, purple-tipped, fragrant flowers, with 5 parts,

are found in upright clusters originating from the leaf axils near the ends of twigs.

Fruit The swollen, elliptic- to kidney-shaped pods, often in pairs, are 10 cm (4 in) long and open when ripe to release numerous flat, brown seeds, each with a parachute of silky, white hairs.

Habitat Although sometimes planted as an ornamental in dry areas, this plant is most frequently seen as an early colonizer of disturbed sites and as an unpalatable relict on overgrazed pastures in dry areas. It is seldom seen growing in open shade and does not occur in mature forests. The absence of other woody weeds growing near this plant indicates that it probably secretes allelopathic compounds.

Purple allamanda   p. 155
Cryptostegia grandiflora   Asclepiadaceae

Distribution The two species in the genus are native to Madagascar. This species was introduced in Mexico two centuries ago by a German sea captain who gave some seeds to a friend in Mazatlán. It was soon widely introduced as an ornamental and has now become naturalized throughout the tropics.

Appearance A sturdy, woody, climbing vine or shrub with arching stems at least 6 m (20 ft) long with an abundant, thick, white, sticky sap present in all parts of the plant except the woody parts of the stem and root.

22

Leaves The dark-green, shiny, leathery leaves are opposite, elliptic and 7 to 9 cm (3 in) long with 5 to 7 mm, yellowish-green petioles and a conspicuous midvein. _Cryptostegia madagascariensis_ has leaves 10 to 12 cm long with reddish 10 to 14 cm petioles.

Flowers The attractive, bell-shaped, pink-purple flowers with a 1.2 cm (0.5 in) calyx are borne in terminal clusters on forked stalks. The similar _Cryptostegia madagascariensis_, with a more reddish hue to the flowers, has a 3 cm (1.2 in) calyx.

Fruit The thick, 3-angled, paired fruits are about 10 cm (4 in) with longitudinal ribs. They split open when mature, releasing the reddish-brown seeds, each with a silky, white parachute.

Habitat It is often planted as an ornamental. It thrives when opportunistically started in sunny openings or open land. It is very drought-tolerant and sometimes runs wild to the proportions of a pest.

Calabash   p. 156
Crescentia cujete   Bignoniaceae

Distribution The genus of 5 species is found in the American tropics. The original range extended from the Bahamas, southern Florida and Mexico south to Peru and Brazil. It has been introduced to southern California, Bermuda, Africa, Asia and some of the Pacific islands.

Appearance A small tree to 10 m (30 ft) tall with sparse, long, spreading, seldom-forked branches. The leaves accentuate the leggy branches by occurring along their full length.

Leaves The oblong leaves, broadest near the end with a prominent midrib, are borne in groups of 2 or more on short spurs arising along the entire length of the branches.

Flowers The light-green, purple-streaked flowers, considered by some to have a fetid odor, are borne on the trunk and along the full length of old branches.

Fruit The hard-shelled, spherical fruits, 10 to 30 cm (4 to 12 in) in diameter, contain thin, brown seeds embedded in a white pulp. A fruit from some selected cultivated trees can have a volume as great as 4 liters (1 gal).

Habitat Found primarily in open woodlands and old pastures.

Wild cinnamon   p. 157
Canella winterana   Canellaceae

Distribution The genus of two species occurs from South Florida south through the West Indies and continental tropical America. This species extends naturally from Florida through the West Indies and has been introduced in Brazil and Venezuela.

Appearance This small tree with an open canopy reaches a height of 10 m (33 ft) with a diameter of 20 cm (8 in). While the outer bark is gray, the inner bark is yellow and aromatic with the smell of mixed spices.

Leaves The glossy, dark-green, spatulate leaves, with smooth edges and inconspicuous veins, grow in whorls at the ends of twigs.

Flowers The dark-purplish-red flowers occur in clusters at branch tips. They produce an abundant nectar heavily used by bees.

Fruit The spherical, red or purplish-black berries have several black, curved seeds.

Habitat The tree may be found on sandy or rocky coastlines and inland in pastures or dry forests at low elevations.

Australian beefwood
Casuarina equisetifolia

p. 158
Casuarinaceae

Distribution This single genus in the family has 25 species found naturally in Australia and adjacent island groups. A native of Southeast Asia and Australia, this species was introduced to the New World as trees were brought to Jamaica in 1788. It has been introduced and naturalized in all tropical lands.

Appearance A tall, open-crowned tree to 40 m (140 ft) tall with a 1 m (3 ft) diameter trunk.

Leaves The leaves are tiny, grayish scales on the green, drooping, flexible, needle-like terminal twigs.

Flowers The tiny female flowers form spherical clusters on the woody twigs, while the male flowers are in cylindrical clusters on the green tips of the twigs.

Fruit The fruit is a warty ball about 15 mm (1/2 in) in diameter, composed of many paired sections which split open at maturity and release a light-brown, winged seed.

Habitat This tree is able to thrive on dry, salty, sandy seacoasts in part because of a root symbiosis with an actinomycete, which allows it to transform atmospheric nitrogen into a form usable by the plant.

Purple queen
Tradescantia pallida

p. 159
Commelinaceae

Distribution The genus of 6 species is native to Texas and Central America. This species from Mexico has escaped from cultivation as an ornamental in many countries in the tropics.

Appearance An intensely violet-purple, erect and clambering, fleshy herb with stems to 40 cm (16 in) tall, sometimes forming dense mats.
Leaves The opposite, oblong leaves to 15 cm (6 in) in length and 2.5 cm (1 in) in width often clasp the stem at the base.
Flowers Borne on an axillary or terminal stem, the flowers emerge from a trough or boat-shaped structure formed by 2 leaves similar to the main leaves but smaller. The funnel-shaped flowers with 3 petals are crimson in the bud, but become a light lavender as they emerge. The conspicuous, yellow, pollen-tipped stamens extend considerably beyond the petals.
Fruit The tiny fruits are both rare and inconspicuous.

Habitat This is an extremely adaptable plant. It is drought-tolerant and forms an open ground cover in partial shade or full sun in many different, challenging soil circumstances.

Oyster lily
Tradescantia spathacea

p. 160
Commelinaceae

Distribution
The genus of 6 species is native to Texas and Central America. This species is native to Mexico, Guatemala and the West Indies. It is widely distributed throughout the tropics by cultivation.

Appearance
A low, perennial herb with a fleshy, reclining or erect stem, to 5 cm (2 in) thick and 20 cm (10 in) long.
Leaves The unstalked, erect, flat, dagger-shaped, brittle leaves to 30 cm (12 in) long are dark-green above and purple below.

Flowers The white flowers are enclosed in a pair of boat-shaped, purple bracts nestled among the leaf bases.

Fruit The rough-textured seeds 3 mm (1/8 in) in length are enclosed in small capsules.

Habitat It is often planted as a ground cover or border plant, but also survives as an epiphyte on exposed rocks, logs and in the forks of large trees.

Wandering jew
Tradescantia zebrina

Balsam apple
Momordica charantia

p. 160
Cucurbitaceae

Distribution The genus of about 45 species is from the Old World tropics, mostly in Africa. This plant has been introduced and naturalized throughout the warm parts of the world. Cultivated forms are grown as annuals or in greenhouses in many temperate parts of the world.

Appearance This pungent-smelling, herbaceous vine with slender, hairy, grooved stems grows rapidly to a length of 12 m (40 ft). It may climb over and densely blanket nearby vegetation.

Leaves The widely spaced leaves with irregularly toothed and notched borders usually have 5 palmate lobes with a finely coiled tendril emerging from the stem at the base of the petiole.

Flowers The bright-yellow flowers on thin stalks have 5 petals and are about 3 cm (1 in) wide.

Fruit The bright-orange fruit of the wild plant has a rough, warty surface and is about 5 cm (2 in) long by 2.5 cm (1 in). Upon ripening fully, the exterior splits along 3 lines and curls open, revealing the bright-red arils covering the flat, brown seeds.

Habitat It grows best on good, moist soil in full sun, but may be found climbing in open forest or over thickets or disturbed sites. Birds are extremely fond of the arils and disperse the seeds widely after feeding on the ripe fruit.

Coontie
Zamia pumila

p. 162
Cycadaceae

Distribution The genus of 15 species commonly called cycads or tree ferns is found from the Bahamas and Mexico south to Bolivia and Brazil. This species occurs in Georgia, Florida, the Bahamas and the Greater Antilles.

Appearance A short, palm-like plant with an upright, rough, cylindrical stem to 40 cm (16 in) tall by 20 cm (8 in) in diameter with up to 12 radiating

leaves. The underground stem is thick and starchy with a far-reaching taproot.

Leaves The stiff, fibrous, fern-like leaves are about 50 cm (20 in) in length and have about 15 pairs of glossy, leathery, opposite leaflets.

Flowers These plants lack traditional flowers and conduct their reproduction by structures that look like cones on male and female plants.

Fruit The thick, angular, orange-to-red seeds up to 25 mm (1 in) long are borne near ground level in a cone covered with coarse, hexagonal scales.

Habitat They may be found in a variety of upland, well-drained sites in the open or under open or dense canopy of larger trees.

Air potato
Dioscorea bulbifera

p. 163
Dioscoreaceae

Distribution The genus of about 200 herbaceous, twining species is found in the warm parts of the world. This species is a native of tropical Asia and Africa, but has been widely introduced in warm parts of the world.

Appearance A thornless, climbing, twining (to the left) aggressive vine to 20 m (65 ft) long with a large, starchy, bitter, underground tuber and abundant, brown, irregularly spherical, aerial tubers borne in leaf axils. Numerous varieties have been developed which bear aerial tubers varying considerably in bitterness, color, flavor and texture. Cultivars emphasize the development of either aerial or underground tubers.

Leaves The alternate, glossy, heart-shaped leaves to 15 cm (6 in) wide have conspicuous characteristic veins and petioles with flaps which encircle the stem.

Flowers The 3 mm (1/8 in), inconspicuous flowers are seldom produced.

Fruit The winged seeds are dispersed by the wind. Although not properly a fruit, the obvious and most effective reproduction is by rooting of the mature aerial tubers which drop from the vine.

Habitat Although it is sometimes cultivated, it is most commonly seen in the wild climbing over fences and trees adjacent to clearings.

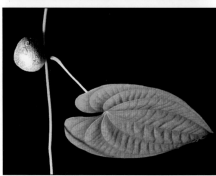

Tung tree   p. 164
Aleurites fordii   Euphorbiaceae

Distribution The genus of five species occurs naturally in Asia and the tropical Pacific. The tung tree has been grown in plantations along the Gulf coast of the U.S. and has been introduced in commercial plantations in the Caribbean and Central and South America. The candlenut tree, *Aleurites moluccana*, is the state tree of Hawaii and has been introduced in South Florida and tropical America as an ornamental. Both species are widely introduced and cultivated in the tropics as a source of shade and oil.

Appearance An open-crowned tree with smooth, gray bark to 12 m (40 ft) tall and wide with a 30 cm (12 in) trunk diameter. It has a milky sap and thick twigs.

Leaves The dark-green, alternate, deciduous, heart-shaped leaves with blades to 15 cm (6 in) long have hairs along the underside of the veins.

Flowers The tree has very attractive, white, rose-throated flowers about 3 cm (1 in) wide borne in terminal clusters composed of both male and female flowers.

Fruit The rounded, pendant-shaped, 5 cm (2 in) fruit of both species has 3 to 7 large, oily, nut-like seeds with a black, brittle, ridged shell. The seeds have an oil content of up to 50%. The unhulled seeds have been mistaken for Brazil nuts and have a flavor which has been compared to chestnuts or toasted almonds.

Habitat The tree seems to thrive on a variety of soil types, but does best on a moist, sandy soil.

Maran   p. 166
Croton astroites   Euphorbiaceae

Distribution The genus of over 600 species of aromatic herbs, shrubs and trees is found worldwide. The species is found from Puerto Rico south to Guadeloupe in the Lesser Antilles.

Appearance A small shrub rarely reaching a height of 5 m (17 ft) with an 8 cm (3 in) trunk. The young branches and leaf petioles are densely coated with orange hairs.

Leaves The pungently aromatic, alternate, elongate-oval-shaped leaves taper to a slender point at the tip. They are 3 to 10 cm (1.2 to 4 in) in length and are covered with star-shaped hairs, white on the leaf blade and orange-rust on the petiole.

Flowers The minute, bell-shaped, greenish female flowers are borne at the base of erect terminal clusters. The male flowers are found at the tips of clusters.

Fruit The oval, 3-lobed seed capsule up to 8 mm (1/3 in) long is supported by the cup-shaped calyx.

Habitat This persistent plant is found on dry hills, disturbed sites and road-sides. It is usually considered a noxious pasture weed.

Broombush   p. 167
Croton betulinus   Euphorbiaceae

Distribution The genus of over 600 species of aromatic herbs, shrubs and trees is found worldwide. This species is found in the West Indies from Cuba to Martinique.

Appearance An open-can-opied, much-branched shrub with the smooth bark of the older stems a dark purplish-brown to black. The stems, both sides of the leaves, and the carpels at the base of the flowers are covered with star-like hairs.

Leaves The ovate leaves, to 3 cm (1 in) long, are borne in alternate clumps of 2 to 5 leaves along the length of the stem.

Flowers The 4-petaled, lavender, trumpet-shaped, 1 cm (0.4 in) wide flower with a pleasant fragrance is borne on a short stalk from the leaf axils near the ends of branches.

Fruit The 1 to 5 (usually 3) seeds are borne in capsules with 5 wings.

Habitat The plant is adaptable and may be found in a dry area with full sun or in the deep shade of the forest.

White maran

Croton discolor

p. 167

Euphorbiaceae

Distribution The genus of over 600 species of aromatic herbs, shrubs and trees is found worldwide. This species occurs in the West Indies.

Appearance An aromatic 2.5 m (8 ft) tall shrub.
Leaves The alternate leaves are distinctly bicolor, with the dark tops contrasting strongly with the lighter undersides.
Flowers The small, inconspicuous male and female flowers are borne on separate stalks.
Fruit The globular seed capsule is about 6 mm (0.25 in) long.

Habitat It is found in open areas, overgrazed pastures and along roadsides.

Wild poinsettia

Euphorbia cyathophora

p. 168

Euphorbiaceae

Distribution This common genus with about 1,600 of species is found worldwide. This species is found worldwide in the tropics and warm temperate regions.

Appearance An annual herb with a smooth, hollow stem, deep taproot, and erect growth habit to 1 m (3 ft).
Leaves The leaves are highly

variable on the same plant, but the leaves near the inflorescence have bright orange-red bases. The very similar *Euphorbia heterophylla* has leaves with white or purple bases.

Flowers The tiny, yellow flowers are borne in terminal groups. The gland on the side of the flower is a flattened oval shape, while the gland on the similar *Euphorbia heterophylla* has a round, slightly flared opening.

Fruit The 3-lobed seed capsule contains 3 gray-brown, 3mm (0.1 in) mottled, rough seeds.

Habitat This plant is most common in disturbed areas and along roadsides.

Candelabra "cactus"
Euphorbia lactea

p. 168
Euphorbiaceae

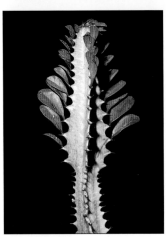

Distribution This common genus with about 1,600 species is found worldwide. A native of India, this plant has been introduced as an ornamental and sometimes naturalized throughout the tropics.

Appearance Usually a shrub with fleshy, 3-angled, dark-green stems with a white to yellow stripe and no conspicuous leaves. It bears paired, sturdy, short, sharp spines on raised teeth along the apex of the longitudinal ridges. It can become a tree to 7 m (22 ft) tall with a 60 cm (2 ft), round, brown trunk. The similar *Euphorbia antiquorum* has 4-angled stems and *Euphorbia royleana* has 5- to 7-angled stems.

Leaves Under some environmental circumstances, this plant produces green or red leaves from the bases of the spines.

Flowers The yellow, cup-like flowers are small, inconspicuous and rare.
Fruit This plant is almost always propagated vegetatively from cuttings, as it seldom sets fruit.

Habitat This drought-adapted plant grows in almost any situation except for salt-saturated soil. It is tolerant of wind and salt spray and has been recommended for coastal sand dune hedges.

Crown of thorns
Euphorbia milii var. *splendens*

p. 169
Euphorbiaceae

Distribution This common genus with about 1,600 species is found worldwide. This species is a native of Madagascar and has been planted worldwide in the tropics.

Appearance A dwarf shrub to 1 m (3 ft) tall with sprawling or erect, spiny branches.
Leaves The oval to pointed leaves are found only near the tips of new growth.
Flowers The inconspicuous flowers on the branch tips have bright-red or pink bracts which resemble petals.
Fruit This plant does not usually set the 3-lobed capsular fruit in our area.

Habitat This plant is tolerant of drought, salt and poor soil but needs full sun. It would be a boon if it produced food.

Black manchineel
Euphorbia petiolaris

p. 170
Euphorbiaceae

Distribution The common genus with about 1,600 species is found worldwide. This is one of the few native trees in the genus and it is found from Hispaniola and the Turks and Caicos islands east into the northern Lesser Antilles and south to Margarita Island.

Appearance A small tree or shrub 6 m (20 ft) tall with a 10 cm (4 in) trunk and shiny, brown, papery, peeling bark.
Leaves The delicate, rounded leaves with 1 cm wide blades and red-tinged petioles longer than the blades are in whorls at the nodes.

Flowers   The tiny, inconspicuous flowers are borne in a greenish cup about 3 mm (1/8 inch) wide.

Fruit   The smooth, rounded, 3-parted seed capsule contains a single egg-shaped, pitted seed in each section.

Habitat   This very attractive, woody plant may be found growing in the understory or in full sun in the dry coastal forest.

Poinsettia

Euphorbia pulcherrima

p. 170

Euphorbiaceae

Distribution   This common genus with about 1,600 species is found worldwide. This species was originally found in Mexico, but is now cultivated worldwide.

Appearance   A shrub to 4 m (13 ft) tall with sparse, thick branches producing a thick, white latex when injured.

Leaves   The alternate or whorled 15 cm (6 in) leaves are elliptical to lance-shaped and may be toothed or lobed.

Flowers   The orange and green, 7 mm (0.3 in) flowers in terminal clusters are greatly accentuated by the surrounding whorl of leaves (bracts), which are usually red but may be pink or white. The maximum color is reached as the days become shorter.

Fruit   The 12 mm (0.5 in) long fruit capsule has a smooth, tan, oblong seed in each of the three lobes.

Habitat It is usually found as an outdoor cultivated ornamental in tropical areas, but is widely distributed as a potted plant in December. It requires a good soil with adequate moisture and is sometimes seen growing wild as a relict or an escapee in tropical areas.

Pencil tree
Euphorbia tirucalli

p. 171
Euphorbiaceae

Distribution This common genus with about 1,600 species is found worldwide. A native of tropical Africa, this tree has been introduced, naturalized and has often become a pest throughout the tropics and adjacent warm parts of the world.

Appearance A shrub which may form a tree to 10 m (30 ft) tall with a 15 cm (6 in) trunk. The foliage is composed primarily of the abundant, cylindrical, succulent, rubbery branches and twigs. Damage to the bark releases an abundant, white, gummy latex.
Leaves The narrow, green, alternate leaves are inconspicuous when present on the pencil-like green twigs which carry on most of the plant's photosynthesis.
Flowers The scarce, seldom-seen flower clusters form ivory-yellow heads 15 mm (5/8 in) across at the ends of branches.
Fruit The 8 mm (0.3 in) long, 3-lobed capsule encloses the 4 mm (0.2 in), smooth, dark-brown seeds.

Habitat An extremely hardy, drought-, and salt-tolerant plant which has been used for coastal sand dune stabilization and as an ornamental in difficult sites. It also withstands the rigors of being confined as an indoor potted plant in northern areas.

Manchineel
Hippomane mancinella

p. 173
Euphorbiaceae

Distribution The single species in the genus is found on both coasts of the warm parts of the Americas, including the Caribbean and the Galapagos Islands. It has been introduced and has become established on the west coast of Africa.

Appearance A small to medium tree reaching a maximum of 20 m (65 ft) in height and 1 m (3 ft) in diameter. Usually it is much smaller. The milky sap

turns black after exposure to air.

Leaves The alternate, oval, green, shiny leaves have 5 to 10 cm (2 to 4 in) blades which are paler and more dull beneath, with finely serrated edges.

Flowers The small and inconspicuous male and female flowers are borne along a green spike at the ends of branches.

Fruit The pleasantly aromatic, 2 cm (1 in) fruit becomes greenish-yellow on ripening and bears a close resemblance to a small apple. The flavor is sweet and mellow.

Habitat The manchineel may be found in a broad variety of habitats. It is most commonly seen growing on sandy sites along the seashore.

Sandbox tree   p. 174
Hura crepitans   Euphorbiaceae

Distribution The genus of 2 species is found in the warm Americas. The original distribution of this tree is from Costa Rica and the Caribbean south to Peru and Brazil. It has been planted in southern Florida, Bermuda, California, Mexico, Africa and Asia and on various Pacific islands.

Appearance A large tree to 30 m (100 ft) tall with a trunk to 1.3 m (4 ft) and a dense, rounded crown. The trunk and branches (particularly of young trees) have many short, sharp, conical spines and contain a clear, sticky sap.

Leaves The oval leaf blades with a short, sharp point are angled up at the midrib and are 15 cm (6 in) long by 10 cm (4 in) broad with petioles of equal length.

Flowers The dark-red, mushroom-shaped female flowers, 3 cm (1 in) long, are borne laterally near branch tips. The smaller, red, male flowers are found in clusters like kernels of corn on the end of a stalk.

Fruit The slightly flattened, round fruit, 8 cm (3 in) in diameter by 3 cm (1 in) thick, has about 15 individually distinct rounded sections, each with a broad, dark-brown seed which reportedly tastes like a walnut. As the fruit dries, each of the segments begins to warp but is restrained by its neighbors. This creates a tremendous balanced internal pressure which continues to build over days and weeks until one segment fails. The fruit then explodes violently and scatters woody fragments and entire seeds in all directions.

Habitat Generally found growing on moist sites with deep soil near watercourses. Although it may be found growing near the seashore, it will not thrive on sand and is only moderately salt-tolerant.

Physicnut

p. 176

Jatropha curcas

Euphorbiaceae

Distribution The genus of 200 species is found worldwide in the tropics. A native of tropical America, this species was under cultivation by Native Americans prior to European colonization. It has been introduced and naturalized worldwide in the tropics and in some subtropical areas such as Bermuda and the Gulf coast of the U.S.

Appearance An open-crowned tree or bush to 8 m (27 ft) tall with a 15 cm (6 in) trunk, spreading branches and thick twigs with leaves in dense clusters at the tips. The plant produces an abundant, sticky, translucent, yellow sap when injured.

Leaves The dull-green, heart-shaped, somewhat lobed, 10 cm (4 in) leaves have yellowish, sunken veins and petioles as long as the blade.

Flowers The yellow-green, 6 mm (1/4 in), bell-shaped, fragrant flowers occur in clusters originating at the leaf bases near the ends of branches.

Fruit The oval, 3 cm (1 in) brown fruit splits into 3 parts when ripe, releasing the 3 oblong, black seeds which have a pleasant, rich but bland flavor.

Habitat This species is often planted as an ornamental and may be found as patches colonizing disturbed sites.

Wild physicnut   p. 178
Jatropha gossypifolia   Euphorbiaceae

Distribution The genus of 200 species is found worldwide in the tropics. This species is native to Central America and the West Indies. It is sometimes cultivated in Florida and the Old World tropics as an ornamental. It has become a naturalized roadside weed in several areas.

Appearance This perennial herb forms a small, spreading shrub with a sparse, open canopy over 1 m (3 ft) in height if left unmolested. It releases a sticky, yellow, translucent sap when injured.

Leaves The alternate, 10 cm (4 in) wide leaves, with hairy margins, are deeply divided into 3 to 5 pointed lobes and may have strong red-to-purple tinges when growing on certain sites.

Flowers The 5-petaled flowers in small, terminal clusters are a deep, rich maroon.

Fruit The oval, 3-lobed, mature, dry fruit is seldom seen because it splits open explosively when dry, scattering the 3 enclosed seeds in all directions.

Habitat This plant grows on cleared land and along roadsides, and may become a major weed in overgrazed pastures. Even goats refuse to eat the foliage of this plant.

Coral tree   p. 179
Jatropha multifida   Euphorbiaceae

Distribution The genus of 200 species is found worldwide in the tropics. This species is a native West Indian recorded from Barbados in the 17th century. It has been introduced and naturalized in the southern U.S. and throughout the tropics.

Appearance A shrub or small tree to 5 m (16 ft) tall with a trunk 8 cm (3 in) in diameter with thick but weak, spreading branches and an open, irregular crown. It yields a sticky, translucent, yellow sap when injured.

Leaves The 20 cm (8 in) wide leaves in clusters at branch tips have up to 11 separated palmate lobes which are further divided and toothed.

Flowers The many small, scarlet, 5-petaled flowers form a flat-topped cluster held above the leaves on an erect, terminal stalk. The similar *Jatropha integerrima* has vivid pink, separate male and female flowers.

Fruit The fruit is a 3 cm (1 in), bright-yellow, slightly fleshy, usually 3-parted capsule containing an elliptic, mottled-brown seed in each section.

Habitat This plant is very adaptable and may be found growing on a windy site with poor, dry, rocky soil in sun or open shade. It forms a more robust plant when provided with good soil and adequate moisture and is widely cultivated as an ornamental.

Cassava   p. 180
Manihot esculenta   Euphorbiaceae

Distribution The genus of about 140 species is native to tropical America. This species is a native of the Brazilian Amazon and has been introduced as a cultivated food-producing plant throughout the warm parts of the world. The plant was widely cultivated in the Americas long before the European invasion. Archaeological excavations have unearthed cassava-processing implements 3,000 years old, and other evidence indicates it has been cultivated for 4,000 years.

Appearance A vigorous, erect, perennial herb 2 m (6 ft) or more tall with single stems thickened at the nodes and branching at flower clusters. The elongate, thick, edible, tuberous roots radiate in clusters at the stem base.

Leaves The alternate, sometimes varie-gated leaves have deeply parted, 3- to 7-lobed blades, 15 cm (6 in) long. The petioles are as long as the blade or longer.

Flowers The green or purple-tinged flowers are borne on a 10 cm (4 in) spike.

Fruit The oval seed capsule is in 3 parts, each containing a mottled seed about 1.2 cm (0.5 in) long.

Habitat This plant often persists and becomes established after cultivation. It is one of the few crops which seem to thrive on nutrient-poor, acid tropical soils.

Christmas candle
Pedilanthus tithymaloides

p. 182
Euphorbiaceae

P. latifolius   P. tithymaloides

Distribution The genus of about 30 species is found in the American tropics. This species occurs naturally from Mexico to Surinam, but is now culti-vated and has escaped throughout the tropics.

Appearance A perennial shrub with many clustered, thick, green, fleshy, erect, zigzag stems to 2 m (6 ft) tall. The plant releases an abundant, milky latex when injured. The similar *Pedilanthus latifolius* has variegated leaves.

Leaves The alternate, nearly stem-less, green or green and white, oblong leaves, often variegated in irregular patterns, are 10 cm (4 in) in length by 5 cm (2 in) in width with a keeled midrib.

Flowers The small flowers are borne in clusters enclosed by red or purple, slip-per-shaped bracts on the branch tips.

Fruit The oval, 3-lobed seed capsule contains mottled, 4 mm (0.15 in), gray-brown seeds.

Habitat It is widely cultivated as an ornamental and may be found growing opportunistically from discarded cuttings. It thrives in hard, dry soils in full sun, but is also shade-tolerant.

Gale-of-wind
Phyllanthus niruri

p. 183
Euphorbiaceae

Distribution The considerably variable genus of over 700 species is most numerous in the Old World tropics. This species is indigenous to the American tropics, but has been distributed worldwide as a weed.

Appearance An erect, annual herb to 50 cm (20 in) tall with 5 to 10 cm (2 to 4 in) horizontal branches.

Leaves The alternate, rounded, elliptic leaves are about 5 to 10 mm (0.2 to 0.4 in) long with very short petioles.

Flowers The tiny male flowers are in groups on a stalk, while the female flowers are solitary on the underside younger branches of the plant.

Fruit The round, smooth seed capsules each contain 6 brown, 1mm (0.04 in) tangerine-segment-shaped seeds with ribbed backs.

Habitat It is found as an early successional weed in grassy areas and on disturbed soil.

Castor bean
Ricinus communis

p. 184
Euphorbiaceae

Distribution The genus contains only this species. A native of the African tropics, it has been introduced as an ornamental shrub and subsequently naturalized ubiquitously in the tropics and subtropics of the world.

Appearance A large, coarse shrub, sometimes becoming a small tree to 10 m (33 ft) in height, with numerous ascending branches, fibrous roots and

a hollow trunk 20 cm (8 in) in diameter. The young stems and leaves are often deep red in color.

Leaves The 30 cm (12 in) petiole joins the 7 to 9 dark-green, palmate lobes in the center of the leaf, defining a blade 40 cm (16 in) or more across. One variety has red or purple foliage.

Flowers The small, ivory-to-rust-colored flowers occur in spikes on the branch tips.

Fruit The 2 cm long, elliptic, brown seed capsules, found in clusters on spikes, are mildly spiny and split explosively into 3 sections when ripe, tossing the 3 shiny, brown, speckled seeds up to 8 meters (25 ft).

Habitat A vigorous colonizer, it may be found on almost any disturbed site.

Mammey apple
Mammea americana

p. 186
Guttiferae

Distribution The genus of about 50 species is native to Africa, Madagascar and Asia with the exception of this tree. A native of the West Indies, this tree has been introduced in Bermuda, South Florida, Bahamas, Central and South America and most other moist, tropical lands.

Appearance A dense, erect tree to 25 m (85 ft) tall with a 60 cm (2 ft) diameter trunk and dense, round crown.

Leaves The glossy, dark-green, alternate leaves up to 20 cm (8 in) long have many parallel veins radiating from the central sunken midrib.

Flowers The fragrant, white, 5 cm (2 in) flowers with 6 petals are borne on twigs or branches behind the terminal leaf clusters.

Fruit The spherical, 8 to 25 cm (3 to 10 in) fruits have a thick, rough, brown, bitter rind over a firm, yellow, juicy, apricot-flavored flesh enclosing 1 to 4 large, oval, reddish-brown seeds.

Habitat The tree thrives best on sites with rich, deep, moist soil. It is quite hardy and survives at a smaller size on many soil types and in areas of only moderate rainfall.

Blood root
Lachnanthes tinctoria

p. 187
Haemodoraceae

Distribution The single species in the genus is found in the southeastern U.S.

Appearance A stout, perennial herb with sword-shaped leaves to 75 cm (30 in) tall. Without fruit or flower, it is superficially similar to cattails, but may be distinguished by crosswise ridges on the leaves which can be felt with a thumbnail. The rhizomes are red and contain a red sap.
Leaves The erect, linear, robust leaves arise from the base of the stem.
Flowers The flowers form erect clusters on a terminal branch.
Fruit The small seeds with woolly parachutes are dispersed by the wind.

Habitat This plant is found in full sunlight in wetlands with seasonally submerged soil.

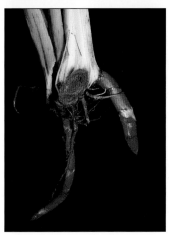

Lion's tail

Leonotis nepetifolia

p. 188

Labiatae

Distribution The genus of 40 species is found in Africa. This species from South Africa is established as a weed in the American tropics and is considered an agricultural pest in many warm, dry parts of the world.

Appearance A sturdy, erect, coarse, annual herb with a square, grooved, hairy, hollow stem to 3 m (10 ft) tall, but usually about 1 m (3 ft). The plant has a long, aggressive, well-developed taproot.
Leaves The opposite, tapered, oval leaves with scalloped edges and conspicuous veins are often finely hairy. The leaves decrease in size upward from the base of the stem.
Flowers The orange-scarlet, tubular flowers form a series of separated, spiny, spherical clusters on the stem.
Fruit The fruit is a ball made up of radiating segments, each of which contains a single, elongate, black-angled seed with a 90° angle on one side and hemicylindrical on the other. As many as six of the spherical fruits may be sequenced on the stem above the foliage.

Habitat It may be found in full sun along roadsides, in overgrazed pastures and on disturbed sites. The seeds germinate rapidly after a rain. The plant grows rapidly and flowers within 5 months, completing its life cycle in 9 months.

Jequirity bean

Abrus precatorius

p. 189

Leguminosae

Distribution The genus of 12 species is found throughout the warm parts of the world. This species is a native of India, but is now found growing wild worldwide in warm climates.

Appearance A slender, twining, climbing vine to 6 m (20 ft) long, developing woody stems in the older parts. The new growth is herbaceous. Stipules often help attach the vine to its supports.

Leaves The alternate, pinnate leaves are 10 cm (4 in) long and have 8 to 15 pairs of leaflets. The leaves droop in the dark.

Flowers The small, attractive, red-to-lavender flowers are borne in clusters on short stalks.

Fruit The flat, oblong pod curls and twists, then splits open when ripe, revealing 3 to 8 almost spherical, shiny, scarlet

seeds, each with a black eye at the point of attachment. The seeds may adhere to the open pods for some time until shaken loose by the wind. A rare form of the plant has white seeds with a black eye and another has a black seed with a white eye.

Habitat It may be seen growing at the edge of clearings or climbing on fences or dead trees in full sun or partial shade.

Angelin
Andira inermis

p. 191
Leguminosae

Distribution The genus of 30 species is found in tropical America, with one species in Africa. This species is found from southern Florida through the Caribbean and Central America to Brazil and Peru. It is also found in tropical West Africa.

Appearance The tree grows to 35 m (120 ft) tall with a thick, sometimes heavily buttressed trunk and a rounded crown. It

becomes more spreading and dense when growing in the open. The ragged, gray bark has a pungent, unpleasant, cabbage-like odor when damaged.

Leaves The leaves are alternate and pinnately compound, 25 cm (10 in) in length, with long, oval, pointed leaflets 8 cm (3 in) long.

Flowers The showy, pink to purple, pea-like flowers are borne in 15 to 30 cm

(6 to 12 in) clusters on the branch tips in early spring of alternate years.
Fruit The abundant fruit is a flattened elliptical pod 3 cm (1 in) in length borne on a stalk. The thin, tough epidermis of the fruit covers a crisp, juicy pulp with a pear-like flavor, which in turn encloses a tough, fibrous husk covering the single seed.

Habitat Although it may be found in many situations due to its nitrogen-fixing roots, it does best in lowland valley bottoms.

Coffee senna
Cassia occidentalis

p. 192
Leguminosae

Distribution The 500 species in the genus are mostly in the warm Americas. This species of American origin has been spread worldwide in the tropics and subtropics.

Appearance This annual or perennial herb is erect, bushy, malodorous and grows to 2 m (6 ft) tall.
Leaves The alternate, 25 cm (10 in) long leaves are pinnately compound with an even number (usually 6 to 12) of pointed leaflets. The leaflets typically decrease in size from the tip to the base of the leaf.
Flowers The yellow-orange, 18 mm (3/4 in) flowers are borne in small groups on short stems originating in the leaf bases or on branch tips, often fading to white and wilting by midday.

Fruit The upright, 10 cm (4 in) long, curved pods are dark-brown in the center and paler on the edges, containing 30 or more dull-brown, oval seeds.

Habitat It is found on disturbed or open sites, along roadsides and in temporarily wet depressions in full sun or very open shade. Soil disturbance seems to promote germination, and the prolific seed production may cause it to become a problem weed in crops and pastures.

Siamese senna

p. 194

Cassia siamea Leguminosae

Distribution The 400 species in the genus are mostly in the warm Americas. A native of the East Indies, this tree has been introduced as an ornamental for its shade and flowers, and as a windbreak. It has become naturalized throughout the tropics.

Appearance A tree with an open, vertical crown reaching 20 m (65 ft) in height with a trunk to 60 cm (2 ft).

Leaves The 25 cm (10 in) long, alternate, pinnate leaves have 12 to 22 pairs of oblong leaflets.

Flowers The abundant, 5-petaled, bright-yellow, 3 cm (1.2 in) flowers are borne as large vertical clusters on the ends of branches.

Fruit The narrow, brown, flat pods, 20 cm (8 in) long, have sequential bulges on alternating sides to accommodate the shiny, flat, dark-brown seeds.

Habitat This tree grows rapidly on a wide variety of moist and dry sites. It is capable of growing on poor soils due to its root nodules, which convert atmospheric nitrogen to usable nutrients.

Shake-shake

p. 195

Crotalaria incana Leguminosae

Distribution The genus of 300 species of herbs with deep taproots is found in warm areas throughout the world. This species originated in the American tropics, but is now found worldwide in the tropics.

Appearance An erect, branching, annual or biennial herb to 1.2 m (4 ft) tall with gray hairs covering the stem.

Leaves The alternate leaves bear 3-parted, nearly round leaflets which arise from a common point. The petiole is longer than the terminal leaflet.

Flowers The yellow to brownish-yellow, 1.2 cm (0.5 in) flowers, sometimes with a reddish tinge, are borne in groups of 5 or more on 10 cm (4 in) terminal stalks.

Fruit The pendulous pod, 3 cm (1 in) in length, is densely covered with fine,

brownish hairs. It splits open at maturity, releasing about 25 hard, green, shiny, kidney-shaped seeds.

Habitat This plant may be found in full sun or partial shade along sandy shorelines, and in other open or disturbed sites which receive moderate to good rainfall.

Yellow lupine
Crotalaria retusa

p. 196
Leguminosae

Distribution The genus of 300 species of herbs with deep taproots is found in warm areas throughout the world. This species is originally from Africa but is now found in the tropics of both the Old and New Worlds.

Appearance An erect, annual or sometimes perennial herb less than 1 m (3 ft) tall with stiff stems.
Leaves The spatulate, alternate leaves have a prominent midvein and very short petiole.
Flowers The yellow to yellowish-red flowers are borne on a terminal stalk.
Fruit The cylindrical, inflated pod, 3 cm (1 in) long by 8 mm (0.3 in) wide, is smooth and hairless with a hooked beak containing many loose, shiny, kidney-shaped, hard seeds.

Habitat It may be found in full sun or open shade in cultivated soil or other disturbed areas such as roadsides or pastures.

Showy crotalaria
Crotalaria spectabilis

p. 197
Leguminosae

Distribution The genus of herbs with deep taproots is found in warm areas throughout the world. This species is a native of Asia but has been widely introduced in the tropics of both the Old and New Worlds.

Appearance An herbaceous annual (sometimes perennial) herb with smooth, erect stems to 2 m (6 ft) tall.
Leaves The oval leaves are up to 15 cm (6 in) long.
Flowers The yellow, pea-like flowers, 25 mm (1 in) long with purple tinges or veins, are found in clusters on branch tips.
Fruit The ripe fruit is an inflated, almost black, beaked pod with about 20 seeds rattling around inside.

Habitat It may be found in open and disturbed areas.

Blue rattleweed
Crotalaria verrucosa

p. 198
Leguminosae

Distribution The genus of 300 species of herbs with deep taproots is found in warm areas throughout the world. This species is found in the Bahamas, West Indies and Central America south to Colombia. It is also found in the Old World tropics.

Appearance An erect, annual herb with zigzag branches to 1 m (3 ft) tall.
Leaves The simple, alternate, wavy-edged leaves, narrow at the base and rounded at the apex, have conspicuous stipules.

Flowers The blue and white, pea-shaped, 1.5 cm (0.5 in) flowers occur in small groups on terminal stalks.

Fruit The cylindrical, hairy, inflated pod with loose, rattling, kidney-shaped seeds is about 3.5 cm (1.5 in) long.

Habitat It is found in old fields and pastures and along roadsides in full sun. The seeds may lie dormant for many years until the soil is tilled or otherwise disturbed.

Dwarf poinciana
Daubentonia punicea

p. 199
Leguminosae

Distribution The genus of 50 species is found in the tropics. This species from Argentina, Uruguay and Brazil is widely introduced, escaped and naturalized in the American tropics and subtropics. It has been declared an exotic weed in South Africa.

Appearance A tree with a bushy, flat-topped canopy to 3 m (10 ft) tall.
Leaves The fine, even, pinnately compound leaves with about 40 leaflets are up to 15 cm (6 in) in length.
Flowers The bright-scarlet, pea-like flowers are borne in elongate clusters from the leaf bases in the summer.
Fruit The 4-winged, red-brown, 10 cm (4 in) pod releases the round, brown seeds when mature.

Habitat
It will grow on many soil types in the sun or full shade. The seeds germinate readily and may form colonies around the parent tree.

Coral bean
Erythrina corallodendrum

p. 199
Leguminosae

Distribution The genus of about 30 species of spiny trees and shrubs is found worldwide in the tropics. This species is a native of the Greater Antilles, but its range has been extended to southern North America and northern South America due to its cultivation as an ornamental. The similar *Erythrina herbaceae* (synonymous with *Erythrina arborea*) is wild in Florida and also widely planted in Florida and the Caribbean.

Appearance A deciduous tree to 25 feet tall with spines on the trunk and branches.
Leaves The compound, alternate, 25 cm (10 in) leaves, with long, sometimes spiny petioles, have 3 wedge-shaped leaflets.
Flowers The blood-red, stalkless flowers are borne in erect clusters, usually when the tree is leafless.
Fruit The scarlet, 9 mm (0.38 in) seeds with a black spot develop in 15 cm (6 in) long pods which are constricted between seeds.

Habitat A component of dry coastal forests or cultivated in yards as an ornamental or as a living fence post.

Madre de cacao
Gliricidia sepium

p. 200
Leguminosae

Distribution The genus of 10 species of trees is found in tropical America. The original distribution of this tree was from Mexico through Central America to the Guianas. It has been introduced and naturalized in the rest of tropical South America, Florida, the West Indies, Africa, Southeast Asia and several Pacific islands.

Appearance A small tree to 12 m (40 ft) in height, with a 50 cm (20 in) trunk and long, slim branches producing an open, irregular, spreading crown.

Leaves The alternate, pinnate leaves, 25 cm (10 in) in length, have 4 to 8 pairs and a single, terminal, 5 cm (2 in), lance-shaped leaflet.

Flowers The abundant, showy, pink- or purple-tinged, pea-like flowers occur as lateral clusters on the branches.

Fruit The flat, yellow-green, 10 cm (4 in) pods turn black and split open when mature, revealing 3 to 8 flat, shiny, reddish-brown to black seeds.

Habitat This tree does well under almost all ecological circumstances except when shaded by larger trees.

Poison indigo
Indigofera suffruticosa

p. 201
Leguminosae

Distribution The genus of about 700 species is found in the warm regions of the world. This species was originally found from the southern U.S. south through South America, but it is now widely naturalized in the Old World tropics.

Appearance A semi-woody shrub to 2 m (6 ft) or more with tall, slender, tough branches.

Leaves The alternate, 10 cm (4 in), unevenly pinnate leaves each bear 9 to 17 1 to 2 cm (0.2 to 0.4 in) oval leaflets.

Flowers The several to many, 3.5 cm (1.5 in), vermilion-to-crimson flowers are borne in clusters on short stalks.

Fruit The flat, sickle-shaped, 2 cm (0.8 in) pod is slightly fuzzy and contains 4 to 8 cylindrical, black seeds.

Habitat It grows as a weed in full sun on cultivated land at the edges of disturbed sites and as a component of brush and scrub thickets. It is still cultivated as a dye plant in Southeast Asia.

Indigofera tinctoria

Wild tamarind

Leucaena leucocephala

p. 202

Leguminosae

Distribution The genus of about 20 species is found in frost-free areas worldwide. The original range of this species is from the Bahamas and Mexico south to northern South America. It has now been introduced and naturalized throughout the tropics.

Appearance A small tree to 7 m (24 ft) tall with tapering, open crown.

Leaves The alternate, twice-pinnate leaves have oblong leaflets lacking petioles. The leaves are often dropped in very dry weather.

Flowers The numerous flowers with short, slender petals and many protruding stamens are massed as white balls 2 cm (1 in) wide borne in clusters on branch tips.

Fruit The fruit is a dark-brown pod 15 cm (6 in) long which splits open at maturity to reveal several dozen shiny, flattened, oblong, dark-brown seeds.

Habitat This aggressive, fast-growing tree thrives in almost any sunny habitat. It does well on soil with a pH range of 5 to 8 and is very drought-resistant. It is considered an unmitigated pest by some and a marvelous source of fuel and livestock feed by others. It is an intermediate in ecological succession in many areas and is eliminated by the shade of large trees.

Jerusalem thorn
Parkinsonia aculeata

p. 204
Leguminosae

Distribution The genus has one species in America and one in Africa. This species is a native of the southwestern U.S. and Mexico. It has been introduced through cultivation and subsequently naturalized from Bermuda south to Argentina and in the Old World tropics.

Appearance A spiny tree with a very open canopy to 6 m (20 ft) tall, with smooth, yellow-green bark and twigs.
Leaves The main axis of the bi-pinnate leaves is short and ends in a spine. The one or two pairs of thin, lateral axes are 25 cm (10 in) long, each with 25 pairs of tiny, deciduous leaflets.
Flowers The showy, bright-yellow, fragrant flowers are borne in open, 15 cm (6 in) clusters on a lateral stem.
Fruit The 2 to 5 dark-brown, oblong seeds form in a 10 cm (4 in) long, brown, papery pod with constrictions between the seeds.

Habitat Although it is a native of the arid West, this tree is adaptable to a great variety of soil conditions, including drought and salt in the full sun.

Cowitch
Stizolobium pruriens

p. 205
Leguminosae

Distribution   The genus of about 120 species is broadly distributed in frost-free areas of the world, with many having stinging hairs on the pod. This species is found from Florida south through the New World tropics and in the Old World tropics. There is considerable confusion on the taxonomy of the cultivated forms of this plant.

Appearance   A slender, twining, fast-growing, aggressive vine climbing to 10 m (33 ft) or more.

Leaves The opposite, 3-parted leaves arise from the vine at 15 cm (6 in) intervals. The central terminal leaflet has a petiole 20 mm in length, while the lateral leaflets have 6 mm petioles.

Flowers The dark-purple, 2.5 cm (1 in) flowers are borne on a short stalk originating from the leaf axil.

Fruit The 8 cm, dark-brown pod is hooked on the lower end, and is covered with over 5,000 fine, stiff, stinging hairs which give it a velvety appearance. The black, lustrous seeds are about 1.2 cm (0.5 in) long.

Habitat It grows in open, sunny places and is often found on disturbed sites such as roadsides, fences and recently cleared land.

Climbing lily
Gloriosa superba

p. 207
Liliaceae

Distribution This genus of 5 species of climbing, tuberous herbs is found in Africa and Asia. This species is from tropical Asia and Africa. It and the similar *Gloriosa rothschildiana* are widely cultivated as ornamentals.

Appearance A modest, perennial vine climbing to 5 m (20 ft) in height. The vine emerges at different places in succeeding years as the tuber, which somewhat resembles a yam, elongates underground. The cooked tuber is juicy, white and somewhat mucilaginous in texture. *Leaves* The dark-green, lanceolate leaves grow to 15 cm (6 in) long. Each has a tendril on the tip which grasps anything within reach.

Flowers The vivid red and yellow flowers with separated, recurved, 10 cm (4 in) long petals turn a uniform red with age. The flowers are used in religious ceremonies in India and Sri Lanka.

Fruit The ribbed, 10 cm (4 in), oblong seedpod splits open at maturity to release the round, scarlet seeds.

Habitat These hardy perennials thrive in a variety of soil types if given full sun and a fence or other support to climb on.

Carolina jessamine

p. 209

Gelsemium sempervirens

Loganiaceae

Distribution The genus of 2 climbing, shrubby vines has one species in North America and one in Asia. This plant is found in the eastern U.S. south through Florida and west to Texas.

Appearance A delicate, climbing, twining (always clockwise) vine with reddish, young stems to 6 m (20 ft).
Leaves The narrow, opposite, 10 cm (4 in), glossy leaves are lanceolate in shape.
Flowers The bright-yellow, fragrant, funnel-shaped flowers are borne in clusters on small, lateral stalks.
Fruit The 12 mm (0.5 in), oval, fruiting capsule splits open at maturity to release its flat, winged seeds.

Habitat This plant is usually found on rich, moist soils growing in the sun or shade.

Mistletoe

p. 210

Phoradendron piperoides

Loranthaceae

Distribution The genus of about 250 species is found primarily in the New World tropics. This species is found in the Greater Antilles, the Bahamas and continental tropical America. There are several similar species with slightly different growth habits and host preferences. *Phoradendron rubrum* is found only on mahogany trees, while *Phoradendron trinervium* is usually found on *Pithecellobium* and *Guapira*. *Phoradendron serotinum* and *flavescens* are very widespread.

Appearance A dense, dark, evergreen shrub which is parasitic on deciduous trees. The roots of mistletoe penetrate the bark and suck water and nutrients from the tissues of the host.
Leaves The opposite, smooth-edged, thick, oval, leathery leaves have a blunt point at the tip.
Flowers The tiny flowers are borne on spikes from the leaf axils. Male and female flowers are on separate plants.

Fruit The fruit is a translucent, fleshy, rounded berry with a small seed embedded in the pulp.

Habitat Mistletoe is only found growing as a parasite on the elevated branches of host trees.

Wild cotton
Gossypium hirsutum

p. 212
Malvaceae

Distribution The 20 species in the genus are spread worldwide. This species originated in Central America and the original distribution of the related species G. *barbadense* was in the West Indies, but both have been widely spread by cultivation and escape. The taxonomic confusion over several species and varieties coupled with extensive hybridization and inaccurate identifications leaves some doubt about exactly which cotton is the subject of many of the chemical and medical reports.

Appearance A shrub or small tree to 4 m (13 ft) feet tall with an 8 cm (3 in) trunk.
Leaves The 5 cm (2 in), 3-to-5 lobed, alternate leaves are oval to heart-shaped with 3 to 5 lobes and long petioles.
Flowers The light-yellow, 7 cm (3 in), bell-shaped flowers are borne singly at leaf bases on sturdy, 1 cm (0.4 in) stalks. The flowers may have red in the throat and turn purple with age.

Fruit The elliptic, 3 cm (1 in) seed capsule (boll) becomes brown and splits open when ripe, revealing a puff of white fibers covering the many seeds. The white, fibrous outgrowths of the seedcoat are the cotton of commerce.

Habitat This plant may be found growing on a variety of sites in the lowlands in full sun or partial shade.

Chinaberry
Melia azedarach

p. 214
Meliaceae

Distribution This genus of 10 species is native to Asia. Originally from southern Asia, this tree has been introduced and naturalized worldwide in the tropics and in temperate climates which have only a few freezing weeks per year. There is considerable confusion in the literature between this tree and _Azadirachta indica,_ which is synonymous with _Melia azadirachta,_ known commonly as the neem tree.

Appearance A medium-sized deciduous tree to 15 m (50 ft) tall with a 60 cm (2 ft) diameter trunk, broadly spreading branches and a flattened crown. It is fast-growing but short-lived. The brittle wood is prone to damage by windstorms.

Leaves The bi-pinnate, alternate, 25 cm (10 in) leaves have toothed leaflets which taper over much of their length to a point and have a distinct odor.

Flowers The tree has showy, branched, 20 cm (8 in) clusters of fragrant, 5-petaled, light-bluish-purple flowers. The flowers are more fragrant at night.

Fruit The 15 mm (0.6 in), spherical, yellow, fleshy, persistent, bitter berries with a hard, 5-seeded stone appear on separate stems in clusters.

Habitat This tree thrives and grows rapidly in a broad variety of habitats and is often planted as an ornamental.

Velvetleaf p. 216
Cissampelos pareira Menispermaceae

Distribution The genus of about 30 species is found around the world in the tropics. This species occurs naturally from Mexico, the Bahamas and the West Indies south to South America; also in tropical Africa and Australia.

Appearance A slender, hairy, perennial, climbing, twining vine with sturdy, twisted, woody roots which are very bitter to taste. The alternate leaves are borne at intervals of about 10 cm (4 in) along the stem.

Leaves The broad, palmately veined leaves are 4 cm (1.5 in) to 10 cm (4 in), and may be heart-shaped, circular or kidney-shaped. The lower and sometimes upper leaf surfaces are velvety due to the presence of soft, white hairs. The petiole is as long as or longer than the leaf.

Flowers The small, yellow male and female flowers are on separate axillary stalks.

Fruit The fruit is a slightly flattened, red or orange, fuzzy sphere about 5 mm (0.2 in) long. The seed is horseshoe-shaped.

Habitat The vine may be found growing naturally on disturbed or early successional sites supported by trees and other growth but seldom on the ground. It does well in sun or open shade, but does not thrive when exposed on windy or salty sites.

Horseradish tree
Moringa oleifera

p. 217
Moringaceae

Distribution The family has only one genus with 10 species found from North Africa to the East Indies. This species is a native of Southeast Asia, but has been introduced as an ornamental and subsequently naturalized worldwide in the tropics.

Appearance A small tree with broad, spreading, drooping, brittle branches to 10 m (30 ft) tall and 25 cm (10 in) in diameter. The thick roots are soft and pungent.

Leaves The thrice-pinnately compound leaves 30 cm (12 in) to 45 cm (18 in) with small, paired, elliptic leaflets give the foliage a feathery, fern-like appearance.

Flowers The abundant, white, fragrant flowers about 2.5 cm (1 in) across occur in clusters on panicles near the ends of branches. Individual trees may flower almost continuously through the year.

Fruit The pointed, pendulous, 3-angled pod is 25 to 35 cm (10 to 14 in) long by 3 cm (1 in) wide. The pod splits open along the angles to reveal linearly arrayed, spherical, brown seeds about 1 cm in diameter, each with 3 equidistant radiating wings.

Habitat This tree is very adaptable and can be seen thriving in full sun on a great variety of soil types and moisture situations. It does best on a well-drained sandy loam with some added fertilizer.

Cajeput
Melaleuca quinquenervia

p. 219
Myrtaceae

Distribution The 100 species of this genus are native to Southeast Asia and the South Pacific. Originally found in Burma, the Malayan peninsula, New Guinea and northern Australia, this species has been introduced and naturalized in many tropical areas.

Appearance A tree to 25 m (85 ft) tall with a narrow, open crown and an irregular trunk with thick, whitish, spongy, peeling bark.

Leaves The stiff, dull-green, alternate, aromatic, 8 cm (3 in) long leaves are narrowly lance-shaped with 3 to 7 longitudinal veins.

Flowers The many small, white flowers are crowded together on branch tips. The masses of erect, thin, ivory stamens produce a result that looks like a bottle-brush.

Fruit The tiny, dust-like, brown seeds are contained in small, rounded capsules crowded on the twigs. The seeds are wind-dispersed within a radius of 15 times the height of the parent tree.

Habitat The tree can grow in many different habitats, but does best in very moist to saturated soil. It is quite resistant to salt, drought and wind and has been widely planted as an ornamental. Introduced to Florida in the early 20th century, it thrives to the point of being considered a noxious, aggressive invader in some areas such as the Everglades of south Florida.

Four o'clock   p. 220
Mirabilis jalapa   Nyctaginaceae

Distribution The genus of 60 species is entirely American in origin. A native of South America, this plant has been introduced worldwide in the tropics and subtropics.

Appearance An erect, perennial herb to 1 m (3 ft) forming a dense, round bush with much spreading and ascending branching. It may develop large tuberous roots to 20 kg (44 lb).

Leaves The opposite, deep-green, narrowly pointed leaves are 10 cm (4 in) long by 5 cm (2 in) wide.

Flowers The fragrant, white, rose, deep-red or purple flowers, about 2 cm (1 in) across, do not open until late afternoon.

Fruit The black, seed-like fruit is covered with tubercles.

Habitat Most commonly seen as a cultivated ornamental, it may also be found growing in open wasteland.

Mexican poppy

Argemone mexicana

p. 221
Papaveraceae

Distribution The dozen species in the genus are found in the warm Americas. This species was originally found in the warm parts of the New World, but was grown from seeds as a garden ornamental in London as early as 1592. It has been introduced and naturalized worldwide.

Appearance A sturdy, coarse, annual herb with a firm tap-root, a single stem and prickly branches, it grows to 1 m (3 ft) tall and has sticky, yellow sap.

Leaves The alternate, blue-green, stalkless leaves are deeply lobed and mottled, with rigid prickles on the margins.

Flowers The yellow, 5 cm (2 in) wide flowers with 4 to 6 petals are found singly on the ends of branches.

Fruit The grooved, spiny, oval, inflated pod with 4 to 6 parts opens at the tip when ripe to release numerous round, black-spotted seeds 2.5 mm (0.1 in) in diameter.

Habitat This plant is found on disturbed sites, on open ground and growing in pastures, usually in full sun, even in areas with little rainfall. Germination of the dormant seeds seems to be stimulated by disturbance and exposure of the soil. It is considered a weed of over 15 different crops in 30 countries.

Love in the mist
Passiflora foetida

p. 224
Passifloraceae

Distribution The genus of over 500 species is found mostly in the American tropics. This Brazilian native species is now introduced and naturalized worldwide in frost-free areas as a weed.

Appearance An herbaceous, bad-smelling, hairy, perennial vine climbing with tendrils to 3 m (10 ft).
Leaves The alternate, 3- to 5-lobed, heart-shaped leaves are sometimes finely toothed on the margins.
Flowers The single, 5 cm (2 in), white and purple flowers, surrounded by 3 frilly bracts, occur in the leaf axils.
Fruit The globose, 3 cm (1.2 in), bright-yellow fruit, with seeds in greenish-yellow pulp, is enclosed in the lacy bracts.

Habitat The vine may be found growing in open areas or climbing over debris or vegetation in any sunny spot.

Pokeweed
Phytolacca americana

p. 224
Phytolaccaceae

Distribution The 35 species in this genus are found in the tropics and temperate regions. This species occurs in the eastern U.S. The very similar *Phytolacca icosandra* is found in the West Indies.

Appearance A leggy, open-canopied herb with coarse, reddish-purple stems to 3 m (10 ft) tall.
Leaves The oval to wedge-shaped, alternate leaves have blades to 20 cm (8 in) with prominent veins beneath.

Flowers The numerous greenish-white to pink flowers are borne on a vertical spike in the variety *rigida*. In the more northern variety *americana*, the spike droops.

Fruit The fruit is a juicy, purple, 7 mm (0.28 in) berry containing about 10 seeds. The seeds are very long-lived and may germinate after as many as 100 years of dormancy.

Habitat This is a native, weedy plant found along fence rows, roadsides, disturbed sites and openings in the forest.

Bloodberry p. 227
Rivina humilis Phytolaccaceae

Distribution The genus of 5 species is found worldwide in warm areas. This species is found from Florida to Texas, in the Caribbean and in tropical America. It has been introduced and naturalized in Asia.

Appearance A bad-smelling, slender-stemmed herb sometimes developing into a shrub up to 1 m (3 ft) tall.

Leaves The soft, lance-shaped, alternate leaves with wavy edges are 10 cm (4 in) long.

Flowers The tiny, white, 4-parted flowers occur on long spikes.

Fruit The glossy, bright-red, 4 mm (0.16 in) fruits with a juicy, orange pulp and a single seed occur in clusters on a spike.

Habitat It is a common weed from the southern U.S. through tropical America and the West Indies.

Sudan grass
Sorghum bicolor

p. 227
Poaceae

Distribution The genus of 30 wild species and about 30 cultivated forms is centered in the Near East. This species from northeast Africa has been cultivated for over 5,000 years and has been developed into many races and varieties as a food and forage plant. It has been cultivated and has become naturalized in the West Indies and other warm parts of the world. There are numerous modern and ancient cultivars taking different names.

Appearance A sturdy, annual grass to 4 m (13 ft) tall without rhizomes and sometimes producing prop roots.
Leaves The ribbon-like, finely-toothed leaf blades with a prominent midrib are up to 60 cm (2 ft) long by 5 cm (2 in) wide.
Flowers The flowers are on slender, terminal stalks.
Fruit The compact seed head with many short branches is up to 40 cm (16 in) long. The seeds of the various varieties may be white, yellow, brown, purple, red or black.

Habitat After cultivation, it often persists and escapes, growing along fence lines, ditches and roadsides.

Johnson grass
Sorghum halepense

p. 229
Poaceae

Distribution The genus of many species is centered in the Near East. This species probably originated in Southern Europe. It was introduced to the U.S. from Turkey in 1830 and has become a weed in many temperate areas of the world.

Appearance A perennial grass to 2 m (6 ft) tall with many scaly, creeping rhizomes.

Leaves The leaf blades, 30 to 90 cm (1 to 3 ft) long and less than 2 cm (0.8 in) wide, have conspicuous, greenish-white midveins.

Flowers The small, inconspicuous flowers are in heads growing from the terminus of the stem.

Fruit The open, loosely-branched, terminal seed head is 10 to 40 cm (4 to 16 in) in length.

Habitat This grass is an aggressive invader and persistent weed. It is found along roadsides, in waste places and sometimes as a weed in other cultivated crops.

Akee   p. 230
Blighia sapida   Sapindaceae

Distribution This genus with a single species is a native of tropical western Africa. It has been widely planted in the American tropics and southern Florida. It is naturalized in Jamaica.

Appearance A medium-sized, open-crowned evergreen tree growing to 20 m (66 ft) tall with stiff branches and a smooth, gray trunk to 50 cm (20 in) in diameter.

Leaves The pinnately compound leaves, more than 30 cm (12 in) long, have 3 to 5 pairs of elliptic, yellow-green, glossy leaflets, each blade 10 to 20 cm (2 to 4 in) long with conspicuous veins.

Flowers The many white, fragrant, 9 mm (3/8 in) flowers are borne on drooping stalks at the base of leaves.

Fruit The fruit is a fleshy, 3-parted capsule 5 to 8 cm in length. When ripe, the capsule develops a red flush and splits open (yawns), revealing 3 large, shiny, dark-brown seeds, each partially encased in a lush white aril. The aril separated from the fruit and the seed is the part prepared as a food.

Habitat It is almost always seen as a yard or garden tree in the New World. It requires full sun and a modest supply of water.

Soapberry tree
Sapindus saponaria

p. 231
Sapindaceae

Distribution The genus of 12 species is found worldwide in the tropics. This species, originally found in tropical America, has been widely introduced throughout the tropics.

Appearance
A tree to 15 m (50 ft) in height with a 30 cm (1 ft) trunk, grooved, flaking bark and a broad crown.

Leaves The 25 cm (10 in), alternate, pinnate leaves have 6 to 12 paired, lance-shaped leaflets with a winged axis and no terminal leaflet.

Flowers The 5 mm (3/16 in), white, 5-petaled flowers are borne in 25 cm (10 in), dense, lateral, branched clusters.

Fruit The shiny, 2 cm (1 in), spherical berries have a waxy, translucent-yellow, bitter, sticky flesh enclosing a round, hard, 9 mm (3/8 in), smooth, black seed. The ripe berries may remain on the tree and dehydrate to become wrinkled and brown.

Habitat This tree does well in relatively dry areas with full sun at low elevations. The tree is also quite salt-tolerant.

Day-blooming jasmine
Cestrum diurnum

p. 232
Solanaceae

Distribution The genus of 150 species is found in the warm Americas. This species is a native of the West Indies and Central America. It has been spread into South Florida and South America by cultivation as an ornamental, with subsequent escape and naturalization.

Appearance Usually a shrub which may become a small tree to 4 m (14 ft) tall with an 8 cm (3 in) trunk.

Leaves The alternate, 10 cm (4 in), oblong leaves are curved at the sides. They are shiny-green above and dull whitish-green below with a pale midrib and 1 cm (3/8 in), light-green petioles.

Flowers The several to many diurnally fragrant, bell-shaped, white flowers with rounded petals are borne in clusters at the bases of the uppermost leaves. Flowering is at least twice a year. Certain bushes under favorable circumstances flower almost continuously.

Fruit The 8 mm (0.3 in), oblong, green berries with several brown, angular seeds progress from white to lavender to purple to black when ripening.

Habitat This plant may be found along roadsides, in pastures or at other sunny, disturbed sites. It is widely dispersed by the seeds in bird droppings.

Night-blooming jasmine

Cestrum nocturnum

p. 233
Solanaceae

Distribution The genus of 150 species is found in the warm Americas. This species is a native of the Greater Antillean islands of Cuba, Jamaica and Hispaniola as well as Central America from Mexico to Honduras. It has been introduced and sometimes naturalized from South Florida to Brazil, on many Pacific islands and in the Old World tropics.

Appearance A bushy shrub with many drooping branches to 4 m (12 ft) tall.

Leaves The glossy, alternate, lanceolate leaves with rounded bases and long, pointed apices vary in length from 7 to 15 cm (3 to 6 in).

Flowers The many fragrant, white, tubular flowers with pointed petals occur in branched clusters at leaf bases and release their abundant fragrance at night.
Fruit The clustered, 9 mm (3/8 in), oblong, green fruits turn white when ripe. They have several dark-brown seeds and a calyx attached to the base.

Habitat This plant thrives on sunny lowland sites.

Angel's trumpet
Datura candida

p. 234
Solanaceae

Distribution The genus of about a dozen species is widely distributed in the warm parts of the world. This species is a native of Peru, but has been widely distributed by cultivation, escape and naturalization. There is some confusion in the literature between this plant and the similar *Datura suaveolens*.

Appearance An open-canopied tree with spreading branches to 5 m (18 ft) tall with a 15 cm (6 in) trunk.
Leaves The long-petioled, drooping, fuzzy leaves are elliptic and alternate and up to 40 cm (18 in) long by 12 cm (5 in) wide.
Flowers The 20 to 30 cm (8 to 12 in) long, fragrant, white (sometimes with a slight peach or pink hue) flowers are trumpet-shaped with a long, pointed calyx and dangle vertically. The flower of the similar *Datura suaveolens* hangs at a slight angle and has a shorter, 5-pointed calyx and light-green stripes in the sometimes purple-tinged petals. The flowers of *Datura arborea* have a deciduous calyx and notches between the petals.
Fruit The spindle-shaped fruit to 20 cm (8 in) in length contains rough, 6 mm (0.25 in) seeds. This plant seldom produces mature fruit.

Habitat It is found in open or disturbed sites, is persistent and escapes after cultivation.

Devil's trumpet
Datura metel

p. 235
Solanaceae

Distribution The genus of about 12 species is widely distributed in the warm parts of the world. This species is a native of tropical Asia, but is now widely introduced and naturalized throughout the warmer parts of the world.

Appearance An erect, coarse, annual herb with a sturdy branching stem to 2 m (6 ft) in height. The entire plant often has a purplish hue.

Leaves The malodorous, alternate, downy, ovate to elliptic leaves are up to 25 cm (10 in) long.
Flowers The erect, white flowers, with the tubular part 10 to 20 cm (4 to 8 in) long, are tinged with violet or purple on the outside.
Fruit The oval fruit has short spines or tubercles and nods on its stem. When ripe, the 4-celled pod splits open irregularly and releases many flattened, kidney-shaped, light-brown, 6 mm (0.24 in) seeds.

Habitat It is most frequent on thin sandy soils, but is so vigorous and adaptable that it may be found as a weed of cultivated crops, in pastures or on almost any disturbed soil in the full sun.

Jimsonweed
Datura stramonium

p. 237
Solanaceae

Distribution The genus of about 12 species is widely distributed in the warm parts of the world. This native of Asia is now distributed widely in tropical and temperate regions of the world.

Appearance An erect, coarse, spreading annual with a sturdy, hollow, fuzzy stem to 1 m (3 ft) tall.
Leaves The alternate, 20 cm (8 in) leaves have many sharply toothed lobes along the margins. The foliage has an unpleasant smell when damaged.
Flowers The dangling white flowers, sometimes tinged purple with anthocyanin, have tubes less than 10 cm (4 in) in length.
Fruit The fruits have short, sharp spines and stand erect until they split evenly into 4 segments, releasing the rough, dark-brown seeds.

Habitat It is found growing as a weed in open or disturbed areas, singly or in small groups, in full sun or partial shade.

Christmas berry

Lycium carolinianum

p. 239
Solanaceae

Distribution The genus of about 85 species is found primarily in the warm Americas. This species is found on the coasts of the southeastern U.S. The similar *Lycium tweedianum* is found from the Bahamas through the West Indies to Argentina.

Appearance A spiny, upright, spreading shrub to 3 m (10 ft) tall.

Leaves The sparse, succulent, spatulate leaves are solitary or in clusters of 2 to 4 at the nodes.

Flowers The lilac to pink or white, 5-petaled, 5 mm (0.2 in) flowers are few and solitary on the stem. The similar *L. tweedianum* has a flower with shorter petals.

Fruit The fruit is a brilliant-red berry perched on a short pedicel at the node of a twig.

Habitat This plant is found in coastal areas on saline soil, on disturbed sites and as a component of dry scrub communities.

Tobacco

Nicotiana tabacum

p. 240
Solanaceae

Distribution The genus of about 60 species is naturally distributed from the Americas across the South Pacific to Australia. This species, native to South America, was cultivated and smoked by Native Americans prior to the European invasion. Within a century of its discovery by Europeans, tobacco was being grown and used in all countries with international trade. In spite of its known harmful effects in all forms, it continues to be a major cash crop in America. *Nicotiana rustica,* also called Turkish tobacco, is also widely grown commercially for use in a hookah, for chewing or as snuff. Tree tobacco, *Nicotiana glauca,* from South America, has been widely introduced and naturalized in the West Indies as an ornamental.

Appearance A thick-stemmed, erect, annual herb to 2 m (6 ft) tall. There are many cultivated varieties and local races of this plant.

Leaves The alternate, elliptic to lanceolate leaves grow to 50 cm (20 in) long and the petioles usually clasp the plant stem.

Flowers The cream-colored, tubular flowers, sometimes tinged with pink or red, are up to 5 cm (2 in) long in dense, terminal clusters.

Fruit The pod is an oval capsule containing brown, ridged, elliptical seeds.

Habitat Tobacco is found growing in cultivation, escaped and spontaneously growing in weedy waste sites and disturbed areas.

Chalice vine
Solandra guttata

p. 242
Solanaceae

Distribution This genus of 10 species of climbing shrubs and vines is found in tropical America. The species is a native of tropical America and was widely distributed as an ornamental by the 18th century.

Appearance A rampant, rambling, sturdy vine or shrub climbing to 6 m (20 ft) with strong and tenacious aerial roots.

Leaves The dark-green, oppo-

site, leathery, elliptical leaves to 25 cm (10 in) long are downy beneath and have prominent veins. The leaves are in rosettes on the branch tips.

Flowers The spectacular, goblet-shaped flowers are white when they open, with purple lines in the interior and a prominent pistil. They turn a rich golden-yellow before they drop. The flowers release much more fragrance at night.

Fruit The rarely developed fruit is a 5 cm (2 in), spherical berry partially enclosed by a calyx. The fruit has been reported as edible.

Habitat It does best on modestly fertile soils growing in partial shade.

Black nightshade
Solanum americanum

p. 243
Solanaceae

Distribution The genus of over 1,700 species varies from herbs to trees and is found worldwide, primarily in the tropics. This species has been introduced and naturalized worldwide from 54°N to 45°S.

Appearance An erect or spreading annual or biennial herb to 1 m (3 ft) tall. This plant is widely noted for having considerable morphological variation. It readily forms polyploids and hybrids; thus, the concept of this species is really a complex of subspecies and forms which vary broadly in their chemical characteristics.

Leaves The sometimes toothed or lobed, alternate, oval, pointed, 10 cm (4 in) long leaves tend to have long petioles.

Flowers The 7 mm (0.3 in), white, 5-pointed flowers, with a central clump of stamens showing bright-yellow anthers, are borne in small, lateral clusters.

Fruit The 6 mm (0.25 in), lustrous, globular, black fruit with up to 50 flattened, pitted seeds is found in drooping clusters adjacent to the stems. There may be as few as 20 or as many as 350 berries per plant.

Habitat This plant is widespread and sometimes weedy in open, disturbed areas, roadsides and fence rows. It is considered to be a weed of 37 crops in 61 countries. It thrives only in moist or irrigated environments and does much better in soils with high nitrogen and phosphorus content. The seeds have a high rate of germination and can survive deep burial for up to 40 years, then germinate when brought to the surface and exposed to light by cultivation.

Italian jasmine
Solanum seaforthianum

p. 245
Solanaceae

Distribution The genus of over 1,700 species varies from herbs to trees and is found worldwide, primarily in the tropics. This plant from Central America and the West Indies has been introduced in Europe and the Far East as an ornamental and has become a weed in Australia.

Appearance A climbing vine to 6 m (20 ft) without prickles.
Leaves The pinnate leaves to 20 cm (8 in) long have 3 to 9 leaflets.
Flowers The showy, blue to purple, star-shaped flowers, with a central tuft of bright-yellow stamens, are borne as open bunches on stems from the leaf axils.
Fruit The fruit is a juicy, scarlet, spherical berry 8 mm (5/16 in) in diameter borne in dangling clusters.

Habitat It is found growing in the wild on moist soil in ravines, or in clearings in the forest. It is grown as an ornamental on pergolas, stone walls or fences.

Sky flower
Duranta repens

p. 245
Verbenaceae

Distribution The genus of about 35 species is found in the warm parts of the Americas. This species is found naturally from Bermuda and the southern U.S. south to Argentina. It has been introduced in many other parts of the world.

Appearance An extremely variable plant which may develop into a vine, a shrub or a small tree 5 meters tall with slim, arching branches and an 8 cm (3 in) trunk with long, slender twigs having single or paired spines at the nodes.
Leaves The numerous opposite leaves with blades 4 cm (1.5 in) in length by 2 cm (1 in) wide are a dull light-green on both surfaces and may be toothed

toward the tip. Some cultivated forms have variegated leaves.

Flowers The showy, tubular, 9 mm (0.4 in) flowers may be blue to lavender, arranged along a curving stalk.

Fruit The 9 mm (0.38 in), glossy, oval, persistent fruits are bright-yellow with a bitter flesh and eight seeds.

Habitat This plant prefers dry or open sites with full sun, but also does well in open shade. It sometimes escapes and spreads after being propagated from seeds and cuttings as an ornamental.

Lantana
Lantana camara

p. 246
Verbenaceae

Distribution The genus of about 150 species is found primarily in the warm Americas with a few members in tropical Africa and Asia. This species is found on the Gulf coast of the U.S. south through the Bahamas, the West Indies, Central America and the warm parts of South America. It has been introduced and naturalized to the extent of becoming a weed in Africa, Australia and southern Asia. The plant shows considerable variation in flowers and in size, shape, color and hairiness of the leaves, leading to many taxonomic subdivisions of the species.

Appearance A very bushy, much-branched, perennial shrub to 3 m (10 ft) tall with hooked prickles on the 4-angled stems.

Leaves The toothed, rough-surfaced, opposite leaves to 10 cm. (4 in.) long have conspicuous veins and are pungently aromatic when crushed.

Flowers The yellow to orange-red flowers occur in attractive, flat-topped, 3 cm (1.2 in), clustered heads. There are varieties with yellow, lavender and pink flowers. The outermost flowers open first and begin to darken from yellow to orange to red. As the opening and darkening continue toward the center of the head, a bull's-eye of colors is produced. The similar *L. involucrata* has lavender flowers.

Fruit The shiny, dark, blue-black, spherical fruits, 3 mm (1/8 in) in diameter with 1 or 2 hard seeds, occur in clusters on branch tips.

Habitat It is widely dispersed in bird droppings and is found growing in full sun on disturbed sites. It may initially form dense, impenetrable thickets, but as ecological succession proceeds it may be shaded out by taller plants. The thickets may provide a refuge for still other pests such as rats, wild pigs and tsetse flies. It is an alternate host, allowing for the residual populations of several arthropod agricultural pests. It is considered a weed of 14 crops in 50 countries and is almost ubiquitous as a pasture weed in warm areas. Lantana is also grown as an ornamental both in pots and in the garden.

Animals

Fire sponge, "Don't touch me" sponge
Tedania ignis, Neofibulia nolitangere

p. 249
Porifera

Tedania ignis
Fire sponge on a conch shell.

William B. Gladfelter

Neofibulia nolitangere

Distribution The genus *Tedania* is found in all oceans, even under the ice in Antarctica. This sponge is found in the warm, clear, shallow waters of the Atlantic and Pacific oceans. The "don't touch me" sponge is also found in Florida and Caribbean waters.

Description The fire sponge is orange to red and is often found encrusting on solid marine objects. It is soft to the touch and easily torn. The "don't touch me" sponge is stiff and almost rocklike to touch with a texture like fine sandpaper, massive to encrusting, brown, brittle and crumbly. It often has short, thick chimneys.

Habitat The fire sponge grows attached to a solid substrate in clear, warm, tropical waters. It may be found growing on the shells of large mollusks such as conch, on coral rubble and on mangrove roots. The "don't touch me" sponge is found on north slopes or under ledges, usually in over 10 m (33 ft) of water.

Reproduction Sperm are released into the water and taken up by other sponges to fertilize their eggs. Sponges often retain their fertilized eggs until the embryos develop into competent larvae. The larvae are expelled into the plankton and eventually settle to the bottom to develop into a new sponge. The "smoking sponges" observed by divers are releasing clouds of sperm. The "smoke" is not related to the "fire" which is induced in the victim's skin after contact with the sponge.

Food Sponges feed by filtering plankton and organic detritus as they draw the surrounding water through the sponge structure. They are noted for being able to remove extremely small particles such as bacteria from the water.

Longevity Sponges are slow-growing and may live more than 10 years.

Feather hydroid
Gymnangium longicauda

p. 250
Hydrozoa

Distribution Hydroids may be found in all waters of the Gulf and Caribbean which support coral growth.

Description This species is an example of a large group of marine organisms generally known as hydroids, which look like groups of feathers or little bushes to about 15 cm (6 in) tall. They typically have a dark-brown or black stalk with whitish branches. Each feather or bush is actually a colony of tiny, individual, tentacled animals called polyps.

Habitat These colonies of polyps attach to a hard substrate such as rock, dead coral, pilings, mooring lines or wrecks. They are usually in clear water with some current.

Reproduction This hydroid extends its colonies from a creeping rhizome and produces clumps of colonies, just as certain grasses produce clumps when multiplying by rhizomes.

Food The polyps feed by filtering floating food in the form of plankton.

Longevity Because of its continuous development of new colonies by growth and division, there is no fixed life span for a clump of feather-like colonies.

Fire coral
Millepora alcicornis

p. 250
Hydrozoa

Distribution This genus is represented by three species, with this one being found in all the tropical oceans of the world.

Description The organism that we call fire coral is actually a colony of hundreds of individual polyps barely visible to the unaided eye. The mustard-yellow-appearing calcium carbonate colony encrusts rocks, sea fan skeletons and other solid marine substrates, including such items as the shells of live conch. When growing erect by itself, it tends to form frilly colonies. *Millepora complanata* forms undulating planes or blades in full sun and *Millepora squarrosa* has vertical plates that join to form a series of boxes, often in shaded areas.

Habitat Fire coral can withstand extreme turbulence and wave action and thus is often the dominant coral in areas subject to heavy wave surge.

Reproduction Although they go through a stage of sexual reproduction which includes free-living forms in the plankton, many colonies reproduce and expand by division of individuals.

Food Fire corals capture planktonic prey with their nematocyst-equipped tentacles.

Longevity Individual colonies can live for hundreds of years if the environmental conditions continue to be suitable.

Portuguese man-o-war
Physalia physalis

p. 251
Hydrozoa

Distribution The Portuguese man-o-war is found throughout the tropical and subtropical oceans of the world.

Description A very distinctive jellyfish with a translucent, bright-blue to azure-purple, balloon-like float up to 30 cm (12 in) long by 15 cm (6 in) high which forms an extended sail. The wind moves the jellyfish awkwardly through the water, leading to it's being named after the old square-rigged Portuguese warships. Some individuals are anatomically suited to sail to the right of the wind and others to the left, but none can tack (go in either direction). The creature we call a jellyfish is actually a colony of hydroids.

Habitat This jellyfish is typically found in the open sea, but onshore winds may bring large numbers up on beaches or into protected bays.

Reproduction Although most jellyfish are the medusae (free-floating) stage of the life cycle, which includes an attached hydra-like polyp, this "individual animal" is actually a colony of individually specialized polyps which provide prey capture, digestive, reproductive and float functions. The polyps each act in some ways as individual organs of the jellyfish "organism."

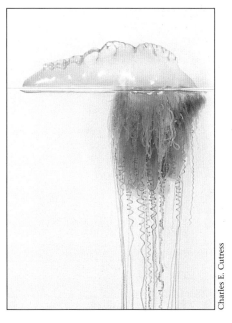

Charles E. Cutress

Food The tentacles, up to 100 feet long, sting and kill small fish and planktonic animals, which are then passed to the mouth and digested.

Longevity The life cycle seems to be completed in one year.

Sea wasp
Carybdea alata

p. 252
Scyphozoa

Distribution This genus is found in the tropical waters of the Atlantic and Pacific oceans.

Description
The jellyfish of this genus are known as cubomedusae or box jelly fish due to their general cubic or rectangular shape when floating freely in the water. They typically have tentacles extending from the 4 corners surrounding the mouth.

Habitat This is an open-water form encountered drifting with the tides. They are attracted to lights at night and are strong swimmers

Reproduction Seasonally, they swarm to the surface to release eggs and sperm.

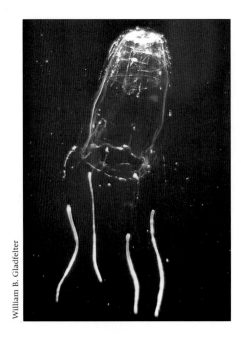

William B. Gladfelter

Food The cubomedusae consumes small fish and other animals which it captures with its stinging tentacles.

Longevity These jellyfish have not been maintained in captivity long enough to determine potential life span, but the life cycle is likely completed in one year.

Upside-down jellyfish
Cassiopeia xamachana

p. 252
Scyphozoa

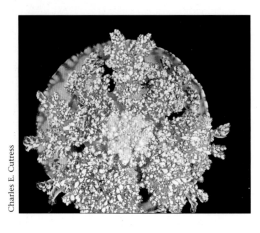

Charles E. Cutress

Distribution The genus of 3 species is found in all of the tropical oceans. This species is found throughout the Caribbean and south Florida.

Description This jellyfish has a disk or bell up to 30 cm (12 in) but is usually smaller. While it is capable of swimming, it typically lays upside down on the bottom. The center of the bell retracts and forms a concavity, allowing it to cling to the substrate. If dislodged from the bottom, it swims with symmetrical contractions of the bell. When at rest, it pulsates gently.

Habitat *Cassiopeia* is found in calm, shallow, clear lagoons or on the shallow, protected sides of reefs and cays.

Reproduction The jellyfish we see on the bottom have sex organs in the frilly arms. When fertilized, the eggs develop into very small larvae called schyphistoma, which attach to the bottom and look like little hydras or anemones. Periodically, these larvae reproduce asexually by budding and producing young medusae stacked like saucers. These break off and after 6 months grow to become the jellyfish we see.

Food While the jellyfish lies with its bell flattened against the bottom, it produces languid pulsations which circulate water over the mouth and arms, providing a source of planktonic food. This is supplemented by the food produced by chlorophyll-containing symbiotic algae growing within the tissues.

Longevity The medusae we commonly see seldom live for more than a year. The small larvae may live for several years.

Sea nettle
Chrysaora quinquecirrha

p. 253
Scyphozoa

Distribution The sea nettle is found worldwide in temperate and tropical waters.

Charles E. Cutress

Description A free-swimming jellyfish with a bell to 18 cm (7 in) with a cluster of lacy central tentacles surrounded by longer, thread-like fishing tentacles up to 1 m (40 in) long. Members of this genus are frequently bioluminescent. A jellyfish population may reach densities of over 50 per cubic meter of water.

Habitat The sea nettle lives in reduced salinity estuaries such as at the mouths of rivers.

Reproduction The early stages of the life cycle occur as red-brown, cupcake-shaped polyps on the underside of solid objects in water less than 11 m (35 ft) deep with salinities reduced to half of seawater or less.

Food This jellyfish seems to feed primarily on small, primarily planktonic animals such as amphipods, protozoa, ctenophores and worms.

Longevity This jellyfish seems to have an annual life cycle related to water temperature.

Giant jellyfish   p. 254
Cyanea capillata   Scyphozoa

Distribution This jellyfish is found in Australia and the North Atlantic and North Pacific oceans.

Description These spherical jellyfish may become so abundant that they clog the nets of fishermen. They are often bioluminescent. Individuals of this genus may grow to be over 2 m (7 ft) across and weigh more than a ton.

Habitat This jellyfish leads a pelagic existence, drifting in the open ocean.

Reproduction
The sessile stage is a yellow-green hydroid 2 mm in diameter shaped like an upside-down pie-pan attached to a hard substrate in reduced salinity estuaries.

Food This jellyfish and its larvae feed on zooplankton.

Longevity The annual life cycle seems to be dependent on water temperature.

Thimble jellyfish   p. 254
Linuche unguiculata   Scyphozoa

Distribution The thimble jellyfish has been found in the tropical Atlantic north through the Bahamas to Bermuda and in the tropical Pacific through the Phillipines to the Maylayan islands. In Florida and the Caribbean it is usually found in the spring and early summer.

Description This jellyfish is only about 12 to 18 mm (0.5 to 0.75 in) in diameter and looks like a short brown cylinder with a few short tentacle around one end.

Habitat This jellyfish is typically found in the open sea but onshore winds may concentrate large numbers near beaches or in to proected bays. They are generally found within 3 m (10 ft) of the surface.

Reproduction It is the microscopic larvae of this jellyfish which cause a problem with swimmers. The larvae may be present in the water in high numbers for short periods of time.

Food The tentacles capture planktonic animals which are ingested and digested.

Longevity The life cycle seems to be completed in one year.

Branching anemone

p. 255

Lebrunea danae

Anthozoa

Distribution This anemone is known from Bermuda to Brazil.

Description *Lebrunea* forms dense carpets on the substrate up to 30 cm (12 in) across.

Habitat The anemone is usually found on a reef, but may be found on a hard bottom.

Reproduction Generations alternate, but the free-swim-

Charles E. Cutress

ming forms are members of the plankton and so are seldom noticed. The attached from that we know as an anemone frequently reproduces by dividing asexually. This gives rise to colonies of anemones which are clones of the originating individual. This type of growth by division also makes possible the huge colonies of *Anthozoa* which we call hard corals.

Food Anemones feed by capturing tiny, swimming planktonic animals and using nematocysts to immobilize them.

Longevity No life history studies have been completed on this anemone, but it seems to live for periods of several years in the same place.

Black sea urchin

p. 255

Diadema antillarum

Echinodermata

Distribution The genus is found in all of the tropical seas of the world. This species is found throughout the warm waters of the western Atlantic from Bermuda to Surinam. It is also found in the tropical waters of the eastern Atlantic.

Description A mature *Diadema* looks like a pincushion with black hat pins. The 5 cm cushion or test is the radially symmetrical body of the urchin, which is protected by a forest of over 100 sharp, mobile, black or purple, brittle spines from 5 to 20 cm in length. Each spine is mounted on a ball joint and is individually controlled by its own musculature, allowing the urchin to selectively point spines in the direction of a disturbance.

Habitat *Diadema* typically cluster by day among coral on reefs or near rock piles. After dark, they venture out to forage on nearby open sand or grass.

Reproduction Sea urchins periodically release masses of eggs and sperm into the water. The fertilized eggs develop into microscopic larvae which drift in the plankton before undergoing metamorphosis and settling on the bottom to develop as new urchins.

Food *Diadema* feed on sea grasses and other vegetable matter. The populations are generally so voracious that they leave a halo of grass-free sand within commuting distance (20 m) of their daytime resting spots in and around reefs and rock piles.

Longevity This sea urchin may live for 5 years or more.

Cone shells
Conus

p. 256
Mollusca

Distribution The 300 to 500 species of *Conus* are found in all the tropical seas of the world. The ones most likely to produce severe symptoms in humans are found in the western Pacific and Indian oceans.

Description The cone shells are generally smooth and conical with long apertures and striking color patterns. The Florida cone, *Conus floridanus,* found in shallow water, has a 12-whorled shell with a high spire which is creamy-white with yellow bands or dotted rows. The agate cone, *Conus ermineus,* is rounded at the shoulder of the spire and is gray-white splotched with chocolate brown. The alphabet cone, *Conus spurius,* has a convex spire and spiral rows of squarish, orange-yellow spots on a creamy-white background. *Conus ermineus* has produced symptoms of toxicity in man.

Habitat *Conus* generally hunt at night in clear, shallow, warm marine environments. By day, they generally take shelter by burrowing into the sand or hiding under rocks and debris.

Reproduction *Conus* males have a penis for internal fertilization of the female. The eggs are deposited in small clumps. When the eggs hatch, a swimming veliger larvae emerges and becomes part of the plankton. When the larvae matures, it settles to the bottom and becomes a small cone.

From left to right: C. spurius, C. Floridanus, C. ermineus

Food *Conus* may feed on mollusks, marine worms or fish. Several cones feed on the polychaete worms called fire worms, which are resistant to many predators due to their irritating bristles. The cones which feed on fish have a large radula and a venom which has the most adverse effect on humans.

Longevity Some species of cones have lived 30 years.

Four-eyed octopus

p. 257

Octopus filosus   Mollusca

Distribution This fascinating genus is found worldwide in tropical and temperate seas. The Gulf of Mexico and the Caribbean have at least 6 species which vary locally in abundance. This species can be identified by a blue ring below each eye, but should not be confused with the blue-ringed octopus of Australia. *Octopus joubini* of the Caribbean has produced human systemic response to its bite.

Description All octopuses have 8 tapering tentacles with suckers on the underside. The tentacles join at a parrot-like beak and a sack-like body. An octopus can change its color, pattern, shape and skin texture at will. An individual may be brown-speckled and knobby, then change to smooth and gray within seconds. They are extremely strong and flexible and can pass through any hole larger than the diameter of the eye.

Habitat They live in holes such as crevices in a reef, conch shells or discarded bottles on the sea floor.

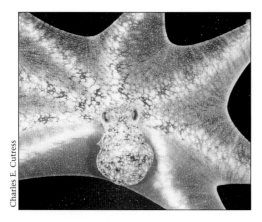

Charles E. Cutress

Reproduction The female deposits her 50,000 eggs in a clump in a den and tends them until they hatch.

Food The octopus usually eats invertebrates such as crabs, snails and clams, but most will catch fish when there is an opportunity.

Longevity An octopus dies after it reproduces at the age of 1 or 2 years.

Red-tipped fireworm
Chloeia viridis

p. 258
Annelida

Charles E. Cutress

Distribution This worm is found in shallow, clear marine waters from Florida and the Bahamas south throughout the Caribbean and on the Pacific coast of tropical Central and South America.

Description A segmented worm elliptical in outline to a maximum of 15 cm (6 in) with a fringe of red-orange tipped white bristles along each side. The caruncle, an appendage on the head, is ribbed and divided along the midline.

Habitat This worm is found in areas with loose bottom debris which offer hiding places from predatory fish.

Food These are fierce predators, consuming anything they can overcome. They eat all forms of animal matter and are sometimes found clinging to the bait in lobster traps and slowly retrieved fish lines.

Longevity This worm may take several years to complete its life cycle.

Orange bristle worm
Eurythoe complanata

p. 258
Annelida

Distribution This ubiquitous marine worm is found in all the tropical seas of the world.

Description A sturdy, orange-yellow, segmented worm, somewhat oval in cross-section to 15 cm (6 in) long, but usually less that half that. The conspicuous gills are red. The worm has a fleshy structure called a caruncle on the head which in this species is elongate and relatively smooth. When the worm is at rest, the bristles are almost concealed in pockets on each side of each segment. If the worm is disturbed, the bristles are extended and splayed so that the worm appears to flower with a mass of white bristles.

Habitat They usually spend the day hiding under rocks or in cracks in the reef, but may rarely be seen wandering about in a manner reminiscent of a centipede. They forage actively in the open at night.

William B. Gladfelter

Darrell Tasman, Clearwater Productions

Reproduction The females may lay thousands of eggs, which complete a complex life cycle as plankton before settling on the reef to mature as worms.

Food These worms feed on the tips of gorgonians.

Longevity Longevity of the worm is unknown.

Green bristle worm, fireworm
Hermodice carunculata

p. 259
Annelida

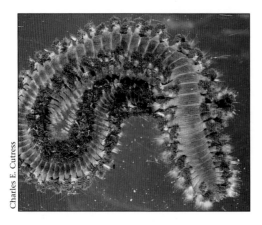

Charles E. Cutress

Distribution This worm is found from the shoreline to water over 100 feet deep throughout the Gulf of Mexico, the Bahamas and the Caribbean.

Description A sturdy, red-green or brown, segmented worm, somewhat square in cross-section, to 30 cm (12 in) long, but usually less that half that. The conspicuous gills are orange. They have a fleshy, branched and pleated structure called a caruncle on the head which is the source of the common name "bearded." In this species, the caruncle is diamond-shaped. When the worm is at rest, the bristles are almost concealed in pockets on each side of each segment. If the worm is disturbed, the bristles are extended and splayed so that the worm appears to flower with a mass of white bristles.

Habitat They usually spend the day hiding under rocks or in cracks in the reef, but may rarely be seen wandering about in a manner reminiscent of a centipede. They forage actively in the open at night.

Reproduction The females may lay over 100,000 eggs, which complete a complex life cycle as plankton.

Food The adults feed on corals, gorgonians, anemones and other attached invertebrates. The feeding of this worm on the fire coral *Millipora complanata* significantly reduces the growth rate of the coral colony and allows colonization by various algae. This worm has also been observed actively feeding on the stony corals *Porites porites* and *P. asteroides*. When this worm feeds on the anemone *Stoichactis helianthus,* the anemone has an unusual defense reaction: it releases its attachment to the bottom, allowing the surf to tumble it about. This frequently detaches the feeding worm. The worm seems immune to the stinging nematocysts on the tentacles of the anemone.

Longevity It is estimated that the adults live several years.

Honey bee
Apis mellifera

p. 260
Hymenoptera

Distribution The genus of 4 species originally occurred only in Africa, Asia and Europe. The honey bee, a native of Europe, now is found in virtually all temperate and tropical parts of the world. The Africanized hybrids introduced in Brazil have recently extended their range through Central America to the southern U.S. and several Caribbean Islands.

Description The honey bee, with a chunky black body and yellow hairy bands, is familiar to all.

Habitat Bees may be found foraging almost anywhere there are flowers. The combs for rearing of brood and storage of honey are formed of wax produced by the bees. The combs are usually placed in cavities but rarely may be initiated in exposed locations.

Reproduction The queen is the sole source of eggs for the maintenance of colony numbers, but the workers raise new queens as needed to produce a continuous egg supply and a colony with an indefinite life span. The social life and functions of the members of a honey bee colony have been the source of several fascinating books.

Food Bees collect the sugar-rich nectar from many species of flowers and concentrate the sugar by evaporating excess water and adding enzymes to produce honey.

Longevity Workers progress through a series of age-specific tasks such as cleaning cells, tending brood, attending queen, packing pollen and guarding the colony entrance. At the age of about 23 days, the workers begin to forage. This task seems to wear them out and cause death in 4 or 5 days. The average worker will fly about 800 km (480 mi) in its lifetime. This distance may be covered in many short flights or fewer long flights in a few days or over a month. The ultimate demise seems to be physical wear on the wings and body as well as the depletion of reserves of glycogen.

North American cow killer
Dasymutilla occidentalis

p. 261
Hymenoptera

Distribution This New World genus of velvet ants has 140 species in North America and many more in Central and South America. This species is found from Connecticut to South Dakota and south to Texas and Florida.

Description The male is winged and does not sting. The female is very attrac-

tive and looks like a large, velvety, orange ant with black bands. She is limited to a terrestrial existence due to a lack of wings. She is well-armed with strong mandibles and a long, agile, probing sting with potent venom. The female has the capability of producing an angry, squeaking stridulation when annoyed. The head, thorax and abdomen are extremely sturdy (over 10 times the strength of those of honeybees) and resistant to the attacks of hosts or the impacts of predators. The wasps have well-developed mandibular glands which can secrete a yellowish, repellent fluid composed primarily of the repellent 4-methyl-3-heptanone.

Habitat These wasps are usually found in open areas moving rapidly about like a frantically foraging ant.

Reproduction The male picks up the female and carries her in flight while mating. Aside from preventing disturbance while conducting this delicate act, the nuptial flight provides a method of dispersal for the earthbound female over barriers such as rivers and swamps. The female spends most of the daylight hours searching frenetically in open areas for potential hosts. She lays her eggs on the larvae of ground-nesting bees and wasps. The velvet ant larvae consume the host larvae upon hatching.

Food The adult wasps feed on nectar and other sweet secretions of plants as well as honeydew produced by aphids.

Longevity These wasps are unusually long-lived. Adult females have been maintained in the laboratory for more than 500 days.

Fire ant   p. 261
Solenopsis geminata   Hymenoptera

Distribution This tramp fire ant probably was introduced to North America from the tropics in the 16th or 17th century. They are now found throughout the southeastern U.S. and the Caribbean. Two other similar species, *Solenopsis invicta* and *Solenopsis richteri*, have been widely established in the Southeast.

Description These medium-sized ants vary in coloration from light reddish to black. They may be seen as dense columns leading from the colony to a food source. Normally observed are the major workers (5 to 6 mm long) with very large heads and the minor workers (2 to 4 mm long).

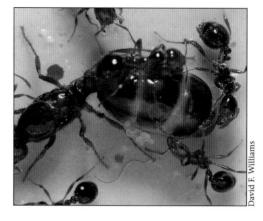

David F. Williams

Habitat The presence of these ants is signaled by large mounds of loose soil.The ant colonies may be found in soil virtually any-where outdoors or around dwellings, usually in sunny locations. They are less common in forests.

Reproduction The queen may lay hundreds of eggs per day. These develop to larvae, pupae and adult workers within a month. New colonies are formed when winged individuals are produced and mate as they disperse.

Food Fire ants prefer to eat animal matter. They kill and eat almost any ter-restrial arthropod they can overpower. The also kill and consume vertebrates which are not able to flee. They will enter the pipped shells of ground nest-ing birds and kill them before they finish hatching.

Longevity While individual ants have a short life span, the life history of the colony allows it to continue for 6 or 7 years.

Paper wasp, hornet, yellow jacket p. 263
Polistes, Vespula, Dolichovespula Hymenoptera

Distribution The 200 species of *Polistes* (paper wasps) found in the New World, and the 25 species of *Vespula* are found worldwide in tropical and temperate areas. These interesting but self-assertive insects are ubiquitous in the southeastern U.S. and the Caribbean.

Description Paper wasps tend to be long, slender and red-brown in color. Hornets and yellowjackets have sturdy, black bodies with conspicuous yellow or white markings.

Habitat The nests are built hanging from the underside of a horizontal sup-port such as a branch or the eaves of a building. Yellowjackets build their paper nests underground and may catch inattentive strollers unawares when

Polistes crinitus

Dolichovespula maculata

the leaf litter near the nest is disturbed by a carelessly placed foot. The adults may be found foraging in many circumstances. Most stings result from disturbance of the nest or from the wasp's concern that damage is imminent.

Reproduction This family of insects produces paper nests by collecting wood fibers and kneading them together with oral secretions. Wasps produce a single horizontal comb of hexagonal cells attached to a support by a pedicel. The hornets may have several combs of cells, all wrapped in a heart-shaped envelope of wood fiber. A single egg is deposited in each downward-facing cell. The adults feed the larvae until it matures and closes the opening of the cell for pupation. Individual cells may be reused, but each queen starts a new colony. While only workers are produced in the spring, reproductive males and females develop in the fall before the colony terminates its annual cycle. In cold climates, the fertilized queens hibernate and start a new colony in the spring. In warm climates, eggs and larvae are present in every month.

Food The adult workers gather nectar as a high-energy food source which they eat themselves. They capture larval and adult insects and spiders to provide a high-protein diet for their larvae.

Longevity The life cycle begins in the spring and is completed by winter.

Yellow fever mosquito
Aedes aegypti

p. 264
Diptera

Distribution About half of the species of mosquitoes in North America belong to this genus. This species is widespread from 40° N to 40° S, but succumbs to temperatures below 8°C or over 37°C.

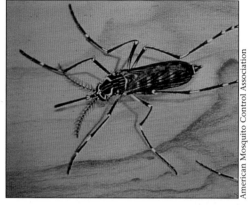

American Mosquito Control Association

Description The adult is conspicuously marked with silver to yellow bands on the black abdomen and similar stripes in a lyre-like pattern on the thorax.

Habitat It does not thrive in high temperatures with low humidity. Other members of this genus are noted for their ability to produce huge numbers of offspring in salt marshes.

Reproduction A female deposits about 140 eggs over a lifetime. The single eggs are deposited by preference at the waterline in small containers of rainwater. They also may breed in the water retained in the leaf axils of bananas and bromeliads or in the brackish water of coastal marshes and swamps. Hatching usually takes about 4 days. The eggs are very resistant to drying and may survive up to a year until rain stimulates hatching. After hatching it takes about a week to complete the growth and 4 molts of the larvae. The robust larvae hang vertically from the water surface. The pupae mature into adult mosquitoes in 1 to 5 days. Under ideal circumstances, the life cycle can be completed in 9 days, but it usually takes about 2 weeks.

Food Both sexes feed on nectar and plant juices for basic metabolic and flight energy. The females feed mostly by day on the blood of many species of mammals, birds, reptiles and amphibians as a source of protein for egg production.

Longevity Females may live a month or more, while males usually survive only about half that time.

Malaria mosquito

Anopheles quadrimaculatus

p. 265

Diptera

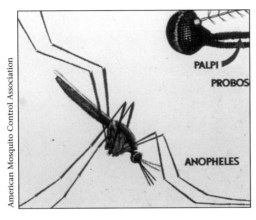

American Mosquito Control Association

PALPI

PROBOS

ANOPHELES

Distribution The genus of over 300 species is found worldwide. There are 90 species in the New World, with about 14 in North America.

Description Anopheles have long legs and few scales on the body. The wings usually have distinct markings. The hum of flight is low-pitched and barely audible when the mosquito flies close to the ear. The most distinct feature is that in feeding, the abdomen, head and body form a straight line, making the mosquito seem to "stand on its head."

Habitat Anopheles are not strong fliers and are most abundant near water sources suitable for breeding. They seldom move more than 1 km (0.6 mi) from the site of their emergence.

Reproduction The females breed shortly after emergence from the pupae and need a blood meal in order to produce eggs. The boat-shaped eggs, with lateral floats, are laid singly in batches of about 200 in a great variety of watery habitats. The preferences of individual species run from brackish water to fresh, from full sun to heavy shade, and from stagnant or trickling water to the water in tree holes and bromeliads. A female may lay 3 or more batches of eggs in her lifetime. The eggs hatch after about 3 days and the larvae develop for 2 weeks before pupating. The larvae are identifiable by their inclination to hang horizontally suspended at the water surface. The adults emerge from the pupae after 30 or so hours.

Food Both sexes feed on nectar and plant juices for basic metabolic and flight energy. The adult females (the only gender to feed on blood) take blood meals from a variety of animals, including man, to provide the protein for egg production. It is this blood-feeding habit which allows the mosquito to transmit diseases such as malaria from one person to another. The males feed on nectar and plant juices.

Longevity Adult females live about a month, and males a considerably less.

House mosquito
Culex pipiens

p. 266
Diptera

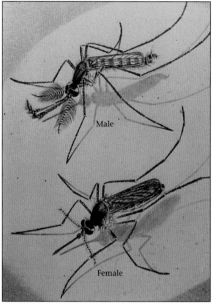

Male

Female

American Mosquito Control Association

Distribution The hundreds of species in this genus are found worldwide.

Description The thorax, abdomen and proboscis of this small mosquito are brown. Its high-pitched whine, vicious bites and sometimes enormous density can make it a terrible pest.

Habitat This mosquito may be found anywhere near the water it needs for reproduction. It is probably the most common species found in homes.

Reproduction After emergence, the female breeds and seeks a blood meal in order to produce eggs. Some populations are able to produce a limited number of eggs without a blood meal. The eggs are laid in groups forming rafts on small bodies of confined water such as rain barrels, catch basins and cisterns. The female may deposit 20 to 30 egg masses, each containing 100 to 200 eggs. In warm summer temperatures the eggs hatch in less than two days, the larvae develop in 10 days or less and the pupae mature in 2 days, yielding a new generation in two weeks or less.

Food Both sexes feed on nectar and plant juices for basic metabolic and flight energy. The females feed on the blood of animals as a source of protein for egg production.

Longevity The adult females live about a month, while the adult males usually live only a week.

Sand flies
Culicoides furens

p. 266
Diptera

Distribution This species is a pest from Massachusetts south to Florida and the Caribbean, including the Pacific coast of Central and South America from Mexico to Ecuador.

Description The light-gray adults are only about 1 mm (0.04 in) long and hold their narrow wings folded, scissor-like, flat over the abdomen when at rest.

Habitat The adults are found within 5 miles of suitable larval habitat, which includes any damp or wet, brackish soil such as that of mangrove swamps.

Reproduction The 0.25 mm (0.01 in), banana-shaped egg is laid on the surface of brackish soil. It hatches within 3 days and completes larval development and pupation to form a new adult within 4 weeks. Females produce eggs within about 2 days of obtaining a blood meal.

Food The females seek a blood meal in order to continue to produce eggs. Feeding is distinctly crepuscular and begins about 40 minutes before sunset and continues for about another 2 hours. A brief rain may induce feeding during daylight hours. Flight is adversely influenced by wind and begins to decline with wind speeds in excess of 2 knots. At 5 knots, sand flies are disinclined to fly. Thus, you can sleep on a breezy Caribbean beach with impunity, but if the wind dies, the sand flies will dine on you with vigor. It is interesting to note that if you are less than 5 meters from dense vegetative cover at feeding time, the flies will walk out to meet you in spite of a brisk breeze. Males do not feed on blood.

Longevity Females usually live about a week after emergence from the pupae, with a maximum life span of about twice that.

Biting spider

Chiricanthium

p. 267

Arachnida

Distribution The genus of 160 species is found worldwide, with at least 6 species having been reported as the source of painful bites. This species was introduced to the U.S. in the early 20th century and has now been spread throughout the country.

Description A small, greenish-white or pale-yellow-tan spider with the cephalothorax darker than the abdomen. The long, delicate legs are quite hairy. The total body length is 7 mm to 16 mm (0.3 to 0.6 in). The bites of these spiders may be quite tenacious, requiring forcible removal.

Habitat The spider does not build insect-catching nets, but uses fine, white, silken threads to build an elongate, sac-like refuge open on both ends. The sacs are often located at the junction of walls with ceilings, on the underside of bookshelves or under eaves.

Reproduction The eggs are deposited in the sac and are tended by the mother until they hatch and disperse.

Food These spiders run and pounce on small insects. They usually are active and feed at night.

Longevity This spider may live for several years.

Black widow spider

Latrodectus mactans

p. 268
Arachnida

Distribution The genus is found worldwide, with at least 5 species in the U.S. This species has been found from Canada to Peru and has been introduced in most warm countries. The taxonomy of the species in this genus is in dispute, with some authors claiming as few as 6 species worldwide and others as many as 29. A detailed analysis using multiple characters and sophisticated techniques will be required to resolve the issue, but there is little doubt which spider we are referring to in this account.

Description A medium-sized, shiny-black spider with long, thin legs and a red hourglass on the bottom side of its globe-like abdomen. The first (front) leg is longer than the fourth (rear). The smaller males do not constitute a public health threat.

Habitat The rather messy, irregular web is built of very tough silk and is often discovered in dimly illuminated, seldom-disturbed areas such as tool sheds, behind shutters or inside buckets. The nest pocket, a tubular retreat in which the spider lurks, is usually centrally located, but may be tucked to one side. In daylight hours, the female prefers to hang inverted.

Reproduction Folklore relates inaccurately that the male is inevitably eaten by the female after mating. The clutch of 100 to 200 eggs is deposited in a silken sac which is tended and moved about by the female until hatching in 2 or 3 weeks. The eggs are often parasitized by fly or wasp larvae. The young are cannibalistic, but soon disperse, often using lengths of web as parachutes to aid in wind transport.

Food The prey consists of any arthropod which blunders into the web, and may include items as large as mice, lizards and grasshoppers.

Longevity The 6 to 8 molts required to reach maturity take 2 to 3 months. The adult female spiders have been recorded as living as long as 849 days, while males rarely survive for 4 months.

Brown recluse spider
Loxosceles reclusa

p. 269
Arachnida

Distribution This genus, with about 80 species, is found in all the warm parts of the world. The thirteen U.S. species in this genus are found primarily in the southern tier of states. This species is found from North Carolina to Texas. It has been moved about widely as a hitch-hiker in household belongings.

Description A medium to small, light-brown spider with a slim build and a violin-shaped pattern on the thorax. It has 3 pairs of pearly-white eyes and a body 8 to 15 mm long.

Habitat When living in houses, it is usually in undisturbed areas such as the backs of closets, basements, storage closets or in seldom-used clothing. It is active at night.

Reproduction It builds a messy, loose, sheet-like web on which it lays its eggs. The females deposit up to 60 eggs in a silken sac. The spiderlings emerge in 25 to 39 days and complete 8 to 11 molts before reaching maturity.

Food This spider eats a great variety of small arthropods. It is reported to be able to live up to 6 months without food.

Longevity The female may live for almost 5 years, while the male rarely exceeds 2 years.

Chigger
Eutrombicula alfreddugesi

p. 270
Arachnida

Distribution This species of chigger is found from Canada through the eastern U.S., Central America, the Caribbean and northern South America.
Description The larvae is a reddish-orange oval, 1.5 mm in length, which may swell on engorgement to as much as 6 mm.

Habitat The distribution is sporadic and localized in all imaginable habitats, from lawns to forests. A heavily infested site may be near an identical habitat which is free of chiggers. Fruit pickers in blackberry thickets are sometimes heavily bitten while friends in another thicket several hundred yards away do not receive a single bite. The mites are active only in warm weather and may be present as little as two months a year in the northern part of the range, while they are active in every month in the deep South. They are most active in moist conditions and may disappear in hot, dry periods.

Reproduction Up to 7 eggs per day per female are deposited in the soil. When the larvae hatch, they crawl to an elevated sprig of vegetation and wait until a suitable vertebrate host appears. After attaching and feeding for up to 3 days on the skin and tissues of the host, the larvae drop off. The larvae metamorphose into 8-legged nymphs which feed on other arthropods before metamorphosing into 8-legged adults. The adults also feed on arthropods, then deposit a new generation of eggs in the soil. Depending on temperature, food and humidity, the life cycle can take from 2 to 12 months.

Food Man is an accidental and very unwilling host of the larvae. They also feed on a wide spectrum of other mammals, birds, reptiles and amphibians.

Longevity It is probably rare for an individual mite to live for more than a year in the wild. Adult females provided with food, water, and warm temperatures in the laboratory have continued to produce eggs after a year. Larvae die if exposed to temperatures below 10° for a lengthy period of time.

American dog tick
Dermacentor variabilis

p. 271
Arachnida

Distribution This genus of about 30 species occurs in Europe, Asia, Africa, and the Americas. This species is widespread in North America and the Caribbean.

Description In the 4th century B.C., Aristotle described ticks as disgusting parasitic animals. Many people still share that view. The ovate, compact body is seed-like, flat, firm, brown, and horny. The body becomes distended, rotund and gray after feeding.

Habitat This tick is commonly found in the highest numbers along trails, paths and openings in the vegetation used for travel by the hosts.

Reproduction The adult tick becomes engorged with blood from her host and drops to the ground. After 3 to 5 days, she deposits about 5,000 eggs in the soil. When the larvae hatch after about 30 days, they climb a wisp of vegetation and attach to the first small rodent which comes along. After feeding for several days, they drop off and molt, becoming nymphs. The larval ticks have only 6 legs, but they develop the adult number of 8 at the nymph stage. The nymphs get on a second host and feed for 3 to 11 days before dropping off to molt into an adult. The adults climb vegetation and wait for a third host animal, to which they attach to feed for 7 to 14 days before mating and dropping off to lay eggs.

Food A blood meal is required by both sexes for each molt from larvae to nymph to adult. The immatures feed on rodents and other small mammals. The adult ticks attach to large mammals such as dogs, horses, and man.

Longevity The life cycle may be completed in as little as 3 months, but ticks are hardy and have a lot of patience. In the absence of suitable hosts, the life cycle may be prolonged. Without feeding, the larvae may live 15 months, the nymphs 20 months and the adults over 2 years.

Scorpion
Centruroides

p. 272
Arachnida

Distribution The genus of 40 species is found from the southern U.S. south to northern South America and the West Indies.

Description At first glimpse, a scorpion appears crab-like, with a flattened body and two conspicuous claws held in front. On closer examination, the long, thin, segmented tail with a stinger (aculeus) on the end positively identifies a scorpion. There are several similar, flattened arthropods with paired claws, but none have a long, 5-jointed tail.

Habitat These scorpions hide by day, but do not burrow. They take shelter by day under rocks, loose bark, petiole stubs of palm trees, or piles of lumber.

They frequently cling to the underside of an object. Fingers wrapped under such an object often squeeze the scorpion, resulting in a retaliatory sting.

Reproduction The male scorpion approaches the female and grasps her claws in his. After a lengthy back-and-forth dance which varies by species, he deposits a spermatophore on the soil which the female inserts in her genitalia. After a gestation period of 10 to 14 months, 7 to 91 scorpions are born alive.. The young may ride on the mother's back for a week or more. They become sexually mature at about 2 years of age.

Food Scorpions come out of hiding at night to search for insects and other invertebrates. They often climb on walls and ceilings in the dark to hunt and eat their prey. Their occasional loss of grip probably accounts for the reports of scorpions dropping on people in bed. Scorpions rip their prey apart and squeeze the fluids from the juicy parts. The victim is stung only when it fails to submit to crushing by the claws and thrashes about excessively while being eaten alive. An offensive sting is delivered with a deliberate thrusting action to a selected soft spot on the victim. Defensive stings are a series of quick flicking thrusts in the general direction of the aggressor. Satiated scorpions tend to be sluggish and mild-mannered, while hungry ones are alert and aggressive.

Longevity Scorpions are slow-growing and may live for many years.

Centipede
Scolopendra subspinipes

p. 273
Myriapoda

Distribution The genus is broadly distributed in the tropics and warm temperate zones. This species is found worldwide in the tropics.

Description Centipedes have the general body form of a large, segmented worm, with a flattened, oval cross-section. A pair of antennae with 17 or more segments protrude from the distinct head. The individual, almost identical segments each have a single pair of legs. Contrary to folklore, the pair of appendages (actually legs) at the rear are totally harmless. This centipede reaches a length of 8 inches. It is fast and very active.

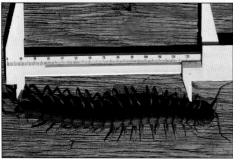

Habitat Centipedes are usually nocturnal and take refuge by day under logs, loose bark, rocks, or debris. They are more common in forested areas, but may be found in almost any habitat which provides cover.

Reproduction The female lays a cluster of 30 to 50 eggs in a brood cavity under a stone or log. The female then coils her body around the eggs and protects them until they hatch.

Food Centipedes feed on insects, spiders, small lizards, snakes, frogs and birds.

Longevity The life cycle is completed in one year, but individuals may live over 3 to 5 years.

Millipede   p. 274
Rhinocricis arboreus   Myriapoda

Distribution This genus of many species is widely distributed in the tropics. This species and several similar ones are found in the Caribbean.

Description Millipedes are cylindrical in cross-section and have many identical body segments, each with 2 pairs of legs. This group of millipedes has only 1 pair of legs on each of the first 5 segments. The antennae are short, with 7 segments. This species is shiny-black and up to 10 cm (4 in) long.

Habitat Millipedes generally prefer to be under cover in damp places, but seasonally, this species roams widely and may be found climbing trees or entering houses and climbing the walls.

Reproduction The small, white eggs, deposited in hidden places, hatch into young millipedes which have only 3 pairs of legs. Additional segments, each with its quota of legs, are added at each subsequent molt.

Food Millipedes are slow and wandering in behavior and are usually vegetarian. Because of weak mouth parts, they prefer to eat decaying plant material or, more rarely, dead insects, mollusks, worms or vertebrates. Circumstantial evidence shows that this large arboreal millipede also eats scale insects.

Note brown stain on finger from secretions

Longevity The life cycle is completed in one year, but the life span may exceed 5 years.

Stingray
Dasyatis americana

p. 275
Fish

Distribution The family of 35 species is found in all the tropical seas of the world. This species is found from Cape Hatteras through the Bahamas, the Gulf of Mexico and the Caribbean to Brazil.

Description Stingrays are cartilaginous fish with flattened, winged, oval to diamond-shaped bodies. They have muscular, highly flexible, whip-like tails with one or more barbed spines.

Darrell Tasman, Clearwater Productions

Habitat Stingrays live on sand or mud bottoms and often cover themselves with sediments while resting.

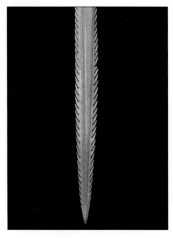

Reproduction This stingray gives live birth to 3 to 5 fully formed young up to 20 cm (8 in) wide.

Food Stingrays eat mollusks, crabs, shrimp, marine worms and other animal matter which they extract from sandy bottoms.

Longevity Little data is available on longevity of stingrays in the wild, but they may live over 10 years in captivity.

Spine from stingray tail

Scorpion fish
Scorpaena plumieri

p. 276
Fish

William B. Gladfelter

Distribution The genus is found worldwide in warm, tropical seas. This species is found on both coasts of the Americas. On the East Coast, it is found from Massachusetts through the Bahamas and the Caribbean to Brazil. There are several other species of scorpion fish with similar appearance and armament in the Gulf and Caribbean.

Description The body of the fish is short and thick, with large scales. The head contains many pits, grooves and fleshy flaps which, combined with the brown- and gray-mottled pattern, produce a great resemblance to a weedy rock. It may reach a length of 17 inches, but is usually only half that size.

Habitat Scorpion fish usually are found lurking near clumps of rocks or in grass in very shallow water.

Reproduction This is one reef fish which does not produce pelagic eggs. They retain the eggs until the young fish are ready to be free-swimming.

Food Scorpion fish sit immobile on the bottom and engulf approaching small fish with astonishing rapidity.

Longevity Scorpion fish have been maintained in aquaria for several years and probably can live much longer than present aquaria records report.

Atlantic puffer, Porcupine fish

Sphoeroides maculatus, Diodontidae

p. 277

Fish

Distribution This family of 100 species and the other families which contain tetrodotoxin are found in all tropical waters.

Description The puffers of this family and the porcupine fishes of the *Diodontidae* both have short, sturdy bodies which they can inflate when threatened by predators. The puffers are smooth and leathery and have small prickles, while the porcupine fish have strong sharp spines erected upon inflation.

Atlantic Puffer

William B. Gladfelter

Habitat Both puffers and porcupine fish occur in shallow, tropical waters on reefs and grass beds.

Reproduction The eggs and larvae are planktonic. Periodically, circumstances prevail which allow a high survival rate of the larvae, leading to large, dense schools of the young fish coming to shore from open ocean waters.

Porcupine Fish

Darrell Tasman, Cleatwater Productions

Food Puffers are carnivorous and eat many different small fish and invertebrates.

Longevity These fish may live more than 10 years.

Scombroid poisoning (tuna and mackerel)
p. 279

Scombridae
Fish

Distribution This family of about 15 genera and 50 species of fish is very important to man both as a primary food fish and as a sport fish. They are found in all oceans of the world; several species of tuna and mackerel are regularly caught in the Gulf of Mexico, the Caribbean and the southwest Atlantic.

Description All members of this family have a row of separated finlets behind the dorsal and anal fins. The body narrows to a small caudal peduncle before flaring to a deeply forked tail. They usually have a metallic coloration and a relatively rich, oily flesh.

Habitat All members of this family are open-water schooling fish. Certain members may approach the shore close enough that fishermen can catch them while standing on the rocks.

Reproduction Mackerel may lay up to 2 million eggs per spawning. These hatch into planktonic larvae having no resemblance to the adult fish.

Food The primary food is small fish, squid and crustaceans, but almost any imaginable animal matter may be eaten at times.

Longevity The life span varies considerably with the size of the species.

Marine toad
p. 280

Bufo marinus
Amphibia

Distribution The genus of about 20 species is found worldwide in tropical and temperate climates. The natural range of this toad is from Texas south to Patagonia. They have been widely introduced both accidentally and intentionally as agents of biological pest control. They are common in Bermuda, on many West Indian islands, and in Hawaii, Australia, on many of the Western Pacific Island groups, in many places in Southeast Asia.

Description This is a very large toad, reaching a body length of 22 cm (8.6 in) and weighing up to 1,400 g (3 lb). It is usually seen about half that size. The body color is generally brown with various intonations of yellow, red, olive, or black. The large parotid gland starting just back of the angle of the jaw is triangular and pitted.

Habitat These toads may be found in almost any habitat within their range. They are nocturnal and most active on drizzly nights. By day, they take shelter under loose boards or piles of vegetation. The may visit pet water dishes regularly to drink and soak if natural water sources are scarce. Rainy nights sometimes produce large aggregations of these toads feeding on insects under street lights.

Reproduction After heavy rains in the spring, the deep, throaty, booming trill of the first arrivals at ponds calls the remaining population out of the hills to what is typically a toad orgy. There is a cacophony of calling with great confusion as both sexes mill about with seemingly random embraces. The males mount and clutch the females as they enter the water. The eggs are fertilized by the males as they are deposited in jelly-covered strings. The black tadpoles hatch after 3 to 4 days and develop within 6 to 7 weeks into tiny toads 12 mm (0.5 in) long which swarm ashore in huge numbers. They reach the minimum breeding size of 9 cm (3.5 in) at 1 year.

Food The primary food is insects, including beetles, caterpillars, ants and cockroaches. They also eat earthworms, snails and mice. Some toads learn to eat pet food and regularly visit dog and cat food dishes left outside.

Longevity Toads are generally long-lived, with several captive individuals having recorded life spans in excess of 30 years. In the wild, predation and natural mortality result in rapid population turnover, with few individuals exceeding 5 years.

Cuban tree frog
p. 282

Osteopilus septentrionalis
Amphibia

Distribution This species is native to Cuba, the nearby Bahamas and the Cayman Islands. It has been introduced to Puerto Rico, the Virgin Islands and southern Florida (including the Keys).

Description This large, nocturnal tree frog with a body length to 14 cm (5 in) is usually a light gray with tints of olive or bronze. The fingers are webbed at the bases and have large, terminal digital disks. The toes are about 2/3 webbed.

Habitat It is widespread in moist habitats, but may become quite abundant in dry areas when moist refuges are provided by man's irrigation or horticultural activities. This frog has the unusual amphibian characteristic of being able to withstand brackish water.

Reproduction The loud, irregular, baying call is given by males only, more frequently after rains as they rest in the branches near small bodies of water. The calls diminish significantly after midnight. The females deposit their clutch of about 130 eggs at the edge of the water while being clasped by the male. The eggs hatch into tadpoles in about a day, then develop into tiny frogs in about 2 months. The tadpoles are aggressive and cannibalistic.

Food The food includes crickets, cockroaches, beetles, moths, caterpillars, flies, spiders, crabs and even other frogs.

Longevity Longevity of the Cuban tree frog is unknown.

Copperhead   p. 282
Agkistrodon contortrix   Reptilia

C. Kenneth Dodd Jr.

Distribution The genus is found in North and Central America, Europe and Asia. Copperheads are found in the southeastern U.S. from Texas to New York, but are common in only 2 counties in northern Florida.

Description The average copperhead is 60 to 90 cm (2 to 3 ft) long with a maximum length of 1.3 m (4 ft 4 in). They have a coppery-brown head and a thick, light-brown to pinkish body with 10 to 20 red-dish-brown, hourglass-shaped crossbands. When coiled and motionless they look astonishingly like a pile of dead leaves.

Habitat They are most commonly found in wooded areas near streams and wetlands.

Reproduction The female retains the eggs until they hatch, then gives birth to live young.

Food Rodents, birds, frogs and insects are all eaten when available.

Longevity The greatest recorded longevity is 29 years, 10 months.

Cottonmouth
Agkistrodon piscivorus

p. 284
Reptilia

Distribution The genus is found in North and Central America, Europe and Asia. This species occurs at low elevations from North Carolina to Texas.

Description A large, dark-olive to black, heavy-bodied snake, usually with 10 to 16 darker crossbands, reaching a maximum length of 1.85 m (6 ft 2 in). The young may have more evident patterns and a bright-yellow tail tip. The white inside of the mouth is conspicuous as a threatening gesture when an aroused snake rears back and holds its mouth open. This snake can be distinguished from the similarly colored and shaped water snakes by the vertical slit pupil and a single row of scales on the underside of the tail. Water snakes have round pupils and a double overlapping row of scales.

Young

Adult

Habitat It is almost always near some form of water, such as rivers, lakes or swamps, and may venture into brackish or salt water. It is active primarily at night.

Reproduction The female retains the eggs until they hatch, then gives birth to live young.

Food This snake has been known to eat fish, frogs, lizards, salamanders, other snakes, small turtles, alligators, birds and mammals. Cottonmouths are preyed upon by large alligators.

Longevity The maximum recorded life span is 18 years, 11 months.

Fer-de-lance
Bothrops species

p. 285

Reptilia

Distribution This genus, with several species in Central and South America, is present on St. Lucia (*Bothrops caribbaea*) and on Martinique (*Bothrops lanceolata*). *Bothrops atrox* is a very common snake in northern South America and Central America.

Description A heavy-bodied snake growing to a maximum of about 2 m (6 ft), but usually considerably less. It is gray to gray-brown with darker markings down the length of the back. A stripe of varying intensity is present on the side of the head.

Habitat It is common along the coast and lowland rivers in rock piles and in piles of debris left by man but may be found hunting in trees. It is primarily nocturnal, but it is often found in shaded areas by day.

Reproduction The 30 to 60 eggs are retained within the female until they hatch.

Food The primary diet is rats and mice, but birds, bats and mongooses are consumed when available. Although the diurnal mongoose seldom encounters this nocturnal snake, the mongoose frequently consumes the snake when they meet.

Longevity The maximum recorded longevity in captivity is 8 years, 6 months.

Diamondback rattlesnake
Crotalus adamanteus

p. 287
Reptilia

Distribution The 30 species of rattlesnakes are found only in the New World from Canada to Argentina. The eastern diamondback is found at low elevations from North Carolina to Mississippi.

Description This rattlesnake is sturdy, with a thick body with a series of dark, diamond-shaped botches. This species is the heaviest of all venomous snakes, reaching a maximum weight of 30 lb. They may reach a length of over 8 feet, but most are half that. The rattle after which this genus of snakes is named is composed of horny, interlocked segments produced at each molting. A rattlesnake may have as many as 23 rattles, although 6 to 10 is more common. Snakes in this family have a pair of deep pits forward of the eyes which can detect small variations in the temperature of the environment. Thus, they can detect and accurately strike a small, warm object such as a rodent in complete darkness. The eye has a vertical-slit pupil with relatively good vision in low light. They do not hear in the conventional sense, but are extremely sensitive to vibrations of the ground.

Habitat This rattlesnake may be found in almost any habitat, except very soggy areas. Highly favored resting places are in the shade of palmettos or next to a log or stump in pine flatwoods. Many more rattlesnakes are encountered than are seen because the coloration tends to break up the outline of the snake, which is often partly concealed by vegetation. Unless overtly threatened, a resting snake will usually remain completely motionless with no rattling. Coastal islands and islands in swamps often have considerably higher populations than those of the adjacent mainland.

Reproduction The 8 to 12 young are born alive from eggs retained in the body of the female until hatching. They are about 35 cm (1 ft) in length when they emerge.

Food This snake has a preference for warm-blooded prey and most frequently feeds on rabbits, squirrels, rats, mice, and birds.

Longevity In the wild, rattlesnakes seldom live longer than 12 years, but in the safety of captivity they may live for 22 years.

Coral snake
Micrurus fulvius

p. 289
Reptilia

Distribution The family is found worldwide and includes notoriously toxic snakes such as cobras, mambas and tiger snakes. The genus has about 68 species from the U.S. south to Argentina. Eastern coral snakes are found at low elevations and along the coast from North Carolina to Texas.

Description The body is patterned, with alternate black and red rings separated by a narrow yellow ring. The tip of the snout is black followed by a yellow band. Similarly colored harmless snakes of several species all have red snouts. Maximum reported size is 51 inches, but most individuals are less than 30 inches long.

Habitat This snake may be seen in almost any natural habitat, from turkey oak and longleaf pine to live oak hammock, and from open scrub to tropical hammock. It is crepuscular to diurnal and seeks shelter in the hot part of the day under decaying vegetation.

Reproduction The 18 cm (7 in) young hatch from the usual 6 or fewer eggs after an incubation of 90 days in moist, rotting logs or other debris.

Food The most common food items are frogs, snakes and lizards.

Longevity The maximum recorded longevity in captivity is 6 years, 10 months.

Pigmy rattlesnake
Sistrurus miliarius

p. 290
Reptilia

Distribution The genus of 3 species is found in the U.S. and Central America. The pigmy rattlesnake is found in the coastal states from North Carolina to Texas.

Description A snake of average build, to a maximum of 31 inches in length, with stippling that may somewhat obscure the 30 to 40 dark, rounded dorsal spots. A reddish-brown dorsal stripe varies in prominence. The sound of the tiny rattle can be easily mistaken for the buzz of an insect. The young have a bright-yellow tip on the tail.

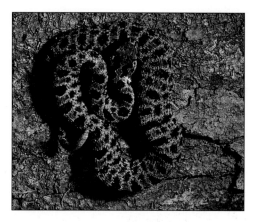

Habitat The pigmy rattlesnake is probably most frequently found in longleaf pine and wire-grass flatwoods, but may be seen anywhere in thickets, or near water in the form of cypress ponds and fresh-water or saltwater marshes.

Reproduction They give birth to living young.

Food This snake eats mice, frogs, lizards, insects and small birds. There are records of them eating centipedes and other small snakes.

Longevity The greatest recorded life span for this snake is 15 years, but they seldom live that long in the wild.

PART II

Toxins, Symptoms and Treatments

Dinoflagellata

Red tide p. 3

Gymnodinium breve Dinoflagellata

Toxic properties The primary toxic component is a group of lipid-soluble polyether neurotoxins called brevitoxins produced by the dinoflagellate *Gymnodinium breve*. The dinoflagellates are filtered from the water as part of the normal feeding process of oysters, coquinas and other clams. The toxin is concentrated in the gut and hepatopancreas and seems to be harmless to the shellfish. Individuals eating the bivalves receive a dose of brevitoxin from the dinoflagellates in the clam's digestive system and perhaps from the toxin which remains in the clams after the dinoflagellates have been digested. The disease produced by the brevitoxin is called neurotoxic shellfish poisoning. *Gymnodinium breve* often can be found in low numbers in coastal waters, but until concentrations approach 5,000 per liter of water, shellfish are seldom sufficiently contaminated to produce symptoms. The amount of toxin produced by individual dinoflagellates seems to vary seasonally. The toxin does not seem to accumulate in or render afflicted fish toxic to human consumption. Although clams may be toxic due to high levels of *Gymnodinium breve* in the water, scallops are not rendered toxic because the muscle is usually the only part eaten.

Symptoms The toxin produces contraction of the smooth muscles of the airways when an aerosol is inhaled. An onshore wind may carry the toxin ashore as part of the sea spray, producing irritation of the eyes, nose, throat and lungs, with subsequent tearing, runny nose, coughing or asthma. The symptoms resolve quite rapidly when the source of irritation is removed by leaving the beach or entering an air-conditioned building. Within three hours (range 15 minutes to 18 hours) after ingestion of shellfish contaminated by brevitoxins, neurotoxic shellfish poisoning may result, with vomiting, abdominal pain, diarrhea, burning pain in the rectum, slowed heartbeat, dilated pupils, and headache. Tingling or numbness of the mouth, throat, trunk or extremities may be accompanied by loss of coordination or reversal of hot-cold temperature sensations. The breakdown of dead dinoflagellates in blooms of red tide releases brevitoxin into the water, resulting in the death of many species of marine fish, of which the most abundant are usually mullet, catfish, and eels. Fish intoxicated by brevitoxin show a twisting, corkscrew swimming motion, with loss of equilibrium before convulsions and death. The toxins may be passed up the food chain to animals other than man. It seems probable that the polyether toxins implicated in the death of many manatees in 1982 on the east coast of Florida originated in the red tide bloom occurring at that time and were concentrated in tunicates eaten by the manatees. A similar series of manatee deaths was caused by a red tide bloom on Florida's southwestern coast in 1996. The high biological oxygen demand resulting from decomposition of dinoflagellates and dead fish may reduce the oxygen level of the water below that needed for survival of many toxin-resistant animals such as sponges, mollusks, crustaceans and worms. The result is that most members of the marine community may die in a severe outbreak of red tide.

Treatment The symptoms are usually self-limiting within less than 2 days. In severe cases, close observation, along with symptomatic and supportive therapy, may be needed. Intravenous fluids may be required to reverse dehydration, and atropine may be indicated to reverse the slowed heartbeat. No human deaths have been reported from eating shellfish exposed to red tide blooms in Florida. Particle filtration masks are usually effective barriers to the inhalation of the airborne toxin.

Beneficial uses The toxin has potential as an experimental tool in biomedical research due to its inhibition of neuromuscular transmission and stimulation of sodium channels in nerves.

Notes The causative organism is also widely referred to in the literature as *Ptychodiscus brevis*. The water-soluble saxitoxin and gonyautoxins produced by the similar dinoflagellate *Gonyaulax* produce a disease called paralytic shellfish poisoning after concentration by filter-feeding mollusks. *Gonyaulax* occurs in the summer in the more northern waters of Europe, along both coasts of the U.S. and in the waters around Japan.

Ciguatera   p. 4
Gambierdiscus toxicus   Dinoflagellata

Toxic properties *Gambierdiscus* is known to produce at least two polyether toxins: lipid-soluble ciguatoxin and water-soluble maitoxin. The toxic dinoflagellates are eaten incidentally by herbivorous fish grazing on algae. The toxin is then concentrated in the food chain and reaches a level harmful to man in large, carnivorous fishes. The toxicity of individual fish is not predictable with reference to species or size, but strictly oceanic fish such as dolphin and tuna whose food chain does not include bottom-dwelling sources do not produce ciguatera. Barracuda are particularly suspect and have been outlawed for sale in several major municipalities such as San Juan, Puerto Rico, and Miami due to ciguatera intoxications. There is no reliable commercially available test for the toxin. Local fishermen are the best source of information on the safety of fish in any area.

Symptoms Ciguatera is manifested as a suite of neurologic, gastrointestinal and cardiovascular symptoms usually evident within a few hours of consuming a contaminated fish. Symptoms may be delayed as much as 24 hours. The neurologic symptoms are the most troubling and may include: numbness and/or tingling of the lips and extremities; itching; trembling; weakness; aching muscles, joints, and teeth; reduced reflexes; and reversal of the senses of hot and cold. Some of the neurological symptoms may persist for months or years. The vomiting, diarrhea and abdominal pains appear early and usually resolve themselves within 24 hours. The heart rate and blood pressure are usually reduced to below normal within 24 hours, but rapid heart rate and dangerously high blood pressure have been reported. The cardiovascular symptoms usually resolve themselves within a few days, but may persist for weeks. There is considerable individual variation in sensitivity to the toxin; several people eating servings of the same fish may show a range of symptoms from asymptomatic to severe incapacitation. The neurological symptoms in general and the itching in particular may continue to reappear years after an exposure upon consumption of alcohol or nonciguatoxic fish. After being poisoned once, a person is usually more sensitive to subsequent exposures, perhaps due to a residue of the toxin in the body or perhaps due to a greater physiological sensitivity to the toxin.

The diagnosis of ciguatera must be based on the circumstantial evidence of consumption of a suspect fish and the immediate onset of consistent symptoms. The disease may be confused with scromboid fish poisoning, neurotoxic and paralytic shellfish poisoning, type E botulism, meningitis, bacterial food poisoning, or poisoning by organophosphate pesticides or monosodium glutamate. Careful investigation of recent food ingestion is the most useful diagnostic tool. Follow-up examination will usually allow the attending physician to discriminate between each of the above afflictions. Because many cases of ciguatera go unreported by residents familiar with the disease, and many others are misdiagnosed by physicians treating tourists after they have returned home to areas where the disease may not be known, the statistics on mortality are uncertain, but have been estimated at between 0.1% and 12.0%. I suspect that actual mortality is lower than the estimate because only the most severe cases reach the attention of medical officials.

Treatment A great variety of treatments have been applied to resolve the symptoms of ciguatera, but none have proven consistently useful in treating moderate intoxications. The administration of a 20% mannitol solution as a piggyback in an IV drip up to a maximum dose of 1 g/kg at the rate of 500 mL/h has brought about dramatic recoveries in severe poisonings. Several folk remedies involve the use of plants containing atropine, and atropine has been used in modern hospital settings to control the vomiting, diarrhea, and depressed pulse rate and blood pressure. A diet free of fish, shellfish and alcohol for 3 to 6 months following an attack is recommended. The above abstinence may need to be lifelong in chronic or repeat intoxications.

Beneficial uses Ciguatoxin has a potential use in biological research, as its mode of action seems to be to increase the sodium permeability of membranes. The investigation of mode of action of the sodium channels in excitable membranes is one of the promising fields in modern biomedical research.

Notes The genus is named for the Gambier Islands in French Polynesia, the source of the first identified specimens. The species name makes reference to the now obvious toxic nature of the dinoflagellate. It is interesting to note that the old Spanish name for the West Indian top shell, *Cittarium pica*, is *Cigua*. This mollusk grazes on the algae growing at and below the wave wash zone of rocky shorelines, the ideal location to find *Gambierdiscus*. Thus, the name for the disease is linked to the shore on which you find *Cigua*. The author contracted a mild case of ciguatera from eating *Cittarium pica* from just such a shoreline. A test to reveal the presence of ciguatoxin in fish flesh would contribute greatly to the well-being of tropical islanders worldwide, not only by preventing medical cases of ciguatera but also by allowing the harvest of many abundant and delectable fish species which are now avoided due to concerns that they may be ciguatoxic. Dinoflagellates which also produce toxins of a similar nature include *Prorocentrum*, *Ostreopsis* and *Coolia*. These may be the sources of toxins in some of the disease outbreaks included under the heading of ciguatera. These multiple potential sources may also explain some of the great variability in the symptoms produced by ciguatera.

Fungi

Mushrooms

p. 5

Amanita and others

Fungi

Toxic properties The primary toxic compound in *Amanita muscaria*, *Amanita pantherina* and several other genera is reported to be muscarine. Ibotenic acid and its decarboxylation derivative muscinol, present in *Amanita muscaria*, kills flies feeding on the mushroom. The cyclopeptide amatoxins in *Amanita phalloides* (and the similar *Amanita tenuifolia*, *Amanita verna* and *Amanita virosa*) are not destroyed by heat and cooking and contribute to the majority of the deaths due to mushroom poisoning worldwide. A fad in the drug culture is the intentional consumption of the hallucinogenic *Psilocybe* mushrooms. The primary toxic component of this group is the compound psilocybin, which affects the central nervous system. A curiosity is the inky cap mushroom, *Coprinus atramentarius,* which is delicious and nontoxic but includes the compound coprine, which slows the metabolism of alcohol.

Symptoms A few minutes to three hours or so after consumption of *A. muscaria*, these symptoms become evident: confusion, blurred vision, constriction of the pupils, salivation, sweating, tearing of the eyes, and severe abdominal pain with frequent, painful, watery diarrhea, followed by slowed heartbeat, blood vessel dilation, and (rarely) coma.

The consumption of *A. phalloides* and relatives allows a delay of 6 to 15 hours before the sudden onset of severely painful abdominal cramps with bloody vomiting and diarrhea. This phase may be fatal unless treated with supportive measures such as fluid and electrolyte replacement in a hospital setting. After one or more days of apparent improvement, liver necrosis and associated symptoms ensue, with an increase in plasma concentrations of liver enzymes and jaundice. Prognosis is poor for patients with thromboplastin (measured blood clotting) time of 10% below normal. Thromboplastin time of 40% above normal usually indicates a full recovery can be expected. Patients may recover slowly after being extremely ill, but improvement may be followed by a fatal relapse.

Within less than an hour after ingestion of *Psilocybe,* the symptoms of euphoria, hallucinations, depression, dilated pupils, nausea, thirst, and frequent, copious and sometimes uncontrolled urination become evident. The symptoms intensify for several hours, then decline and vanish within a day. As with many psychoactive substances, abuse has led to behavior which has resulted in death. Consumption of any alcohol for up to 24 hours after eating the *Coprinus* mushrooms results in the accumulation of acetaldehyde in the body, which in turn produces symptoms very similar to those of disulfiram (Antabuse), with flushing, vomiting, vertigo, palpitations, and other unpleasant feelings for several hours.

Treatment In all cases of mushroom poisoning, the residue of mushrooms should be removed from the gastrointestinal tract as soon as possible to reduce the amount of toxin absorbed. Atropine is the specific antidote for muscarine poisoning in severe cases and provides a good prognosis. Recovery is generally quite rapid and deaths are rare. Modern medicine has developed several marginally effective forms of intervention for the treatment for intoxication by *A. phalloides* and relatives. Thioctic acid has been reported as having antidotal properties to *A. phalloides* poisoning in animal experiments, and limited human trials have shown favorable results. Due to the progress being made in treatment protocols, a current search of the literature via Medline (a computer database of medical literature searchable via modem) or other means is essential in deciding the treatment to apply.

The eventual outcome seems to be strongly related to the amount of mushrooms consumed. Several hundred grams will almost inevitably be fatal, while very small amounts will usually allow eventual recovery with most therapies. Some evidence indicates that the prognosis is improved if patients consumed alcoholic beverages in the same meal with the mushrooms. Caution: In cases of mushroom poisoning, every attempt should be made to obtain and identify a sample of the offending fungus. A poison control center should be contacted for the most current recommended treatment.

Beneficial uses The common, commercially available mushroom is grown under carefully controlled conditions, and safely provides an enjoyable supplement to the diet of millions of people without adverse incidents. Newly developed culture techniques increasingly allow the controlled cultivation of several types of safe, exotic mushrooms increasingly seen in markets. As regards field harvest of this delectable fungus:

There are old mushroom hunters,
and bold mushroom hunters,
but there are no old, bold mushroom hunters.

Notes Although mushrooms are extremely toxic to man, rabbits and squirrels are able to nibble small amounts with impunity, while slugs can exist on a diet composed solely of *A. phalloides*. All mushrooms vary their growth habit depending on circumstances, so a toxic species may look very much like a harmless species. This phenomenon is particularly common in the genus *Amanita*. In chatting with a mycologist who had studied wild mushrooms for 40 years, I asked which ones he preferred to eat. The response was that his identification had been fooled by growth habit so many times that he ate only commercially produced mushrooms.

Vascular Plants

Sisal p. 6
Agave sisalana Agavaceae

Toxic properties The sap from agave leaves produces an irritant contact dermatitis. Calcium oxalate needles in the sap are thought to be the cause of irritation and swelling of the tissues of the mouth and throat when ingested. The juice and pulp from the leaves contain the steroidal sapogenins hecogenin, smilagenin and tigogenin.

Symptoms Skin contact with the sap from the cut leaves produces immediate burning, redness, itching and swelling, followed in several hours by blistering, which heals within 2 weeks. Irritation and temporary blindness result from sap getting into the eye. Infection and inflammation of puncture wounds caused by the sharp leaf tips near bone have resulted in a granuloma reaction resembling a neoplasm. Respiratory allergies have been provoked by the use of sisal fiber as a padding or filler in upholstery and mattresses. The juice from the leaves has been lethal when fed to cattle and rabbits.

Treatment When handling the freshly cut leaves, wear gloves or wash off the sap as soon as possible. When skin is punctured by the leaf tip, be sure that all fragments are removed from the wound, then rinse the wound with hydrogen peroxide. Seek medical care if signs of irritation or infection continue.

Beneficial uses This plant is cultivated extensively in several parts of the tropics for the strong, coarse fibers in its leaves. The fibers of this *Agave* are called sisal, henequin or ixtle and are used to make brushes, rope, twine, sacks, sandals and hammocks. The solid waste from fiber production can be used to make a very strong paper, and a very hard wax similar to carnauba may be extracted. The fluid waste contains pectins used in the food and cosmetic industry. The saponins in the root are sometimes used in washing clothes. A volatile oil from the leaves and seeds has potential for use in controlling mosquito larvae. The juice from the leaves, impregnated in wallpaper, makes the wallpaper unpalatable to termites. Folk medicine has used the juice squeezed from the leaf to treat jaundice, leprosy, syphilis, toothaches, as a purgative or diuretic, and to induce menstrual flow. A concentrated syrup made from the sap is widely used as a folk remedy for treatment of wounds. This folk use is supported by experimental results which show the juice to strongly inhibit *Staphylococcus aureus*, one of the major pus-forming pathogenic bacteria. Hecogenin derived from the waste of sisal fiber production is a source of cortisone and the raw material for the synthesis of oral contraceptives and other steroids. The central stem is sometimes eaten after lengthy, slow roasting. The sap from the flower stalk of *Agave americana* is fermented to produce the beverage pulque, which may be distilled to produce mescal. The pounded leaves have been used to intoxicate fish for easy capture. The copious nectar flow is eagerly gathered by swarms of bees, who produce from it a dark, strong-flavored honey which requires a long ripening time. Hummingbirds and other species of birds feed on the nectar when it is in full flow.

Notes The generic name is from the Greek word *agavos*, "noble." The species name is from Sisal, a place on the coast of Yucatan. The agaves are known by several often-interchanged common names, including century plant and maguey. The early Spanish explorers who found the Native Americans using this plant carried it back and introduced it to southern Europe by the middle of the 16th century.

White atamasco lily
Zephyranthes atamasco

p. 7

Amaryllidaceae

Toxic properties Members of this genus contain the toxic alkaloid lycorine, an emetic with an LD_{50} of 41 mg/kg in dogs. Haemanthidine, widespread in this family, is also present along with nerinine and tazettin. The bulb contains a higher concentration of alkaloids than the foliage.

Symptoms Human ingestion of the bulb produces persistent vomiting and diarrhea. Chronic consumption by experimental animals has produced symptoms similar to scurvy. This plant is suspected of causing the livestock disease called "staggers" in which vomiting, bloody diarrhea and staggering are followed by collapse and sometimes death. The lethal dose of bulbs for steers has been as little as 0.5% of their body weight. Horses and poultry have also reportedly been intoxicated by this plant.

Treatment Fluid replacement and symptomatic support are the only known treatments.

Beneficial uses These lilies are widely planted as perennial ornamentals. Members of this genus have provided several compounds, such as trans-dihydronarciclasine from *Zephyranthes candida*, experimentally active against cancer. Native Americans are reported to have eaten the cooked bulbs in times of food scarcity. The Chinese have used *Zephyranthes* to treat breast cancer, convulsions and hepatitis. Experimentally, the alkaloid lycorine has been found to be a respiratory stimulant and a plant growth inhibitor, and to be active against viruses. Folk medicine has used the bulbs of this genus to treat cancers since the 4th century B.C. A tea made from a dozen leaves boiled in a pint of water is reputed to be a treatment for diabetes if several ounces are taken after each meal. The bulb has also been used to treat tuberculosis and tumors.

Notes The genus is from the Greek *zephyr*, "south wind," and *anthos*, "flower." Atamasco seems to be an Amerindian word. Zephyr flower is another common name for this plant. The flowers often bloom after a substantial rain and are thus are also known as rain lilies. The plant contains nerinine.

Cashew   p. 7
Anacardium occidentale   Anacardiaceae

Toxic properties The shell of the nut contains a caustic, poisonous mixture of phenols including cardol, cardanol and anacardic acid. The active chemicals are very similar in chemical structure to the toxins of poison ivy. The phenols make up what is called the cashew nut shell liquid (CNSL), which must be driven off by roasting before the nuts are cracked or eaten. CNSL, recovered by solvent extraction, is about 90% anacardic acid and 10% cardol. The latter is principally responsible for the irritant contact dermatitis. In CNSL obtained by roasting, the anacardic acid is changed to cardanol. Persons sensitized to poison ivy often are more sensitive to the oil. The sap of the tree is irritating and may induce blisters. The leaves contain hydroxybezoic acid and other polyphenols, along with flavonoid heteromonosides.

Symptoms Direct contact with the shell oil causes severe dermatitis, blistering and swelling. The fumes from roasting can cause grievous injuries to the skin, eyes and respiratory tract on contact. People cutting the wood or even stirring a drink with a cashew stick may be subject to dermatitis. In World War II, people working with aircraft engine ignition insulators or signal corps equipment made from CNSL commonly experienced dermatitis. Modern products are usually more completely polymerized and thus show no irritant properties. Internally, the CNSL produces digestive tract disturbances and, in extreme cases, loss of muscular control and interrupted respiration. Outbreaks of dermatological symptoms have resulted from eating cashew kernels which have been packaged with fragments of shell or contaminated with shell oil.

Treatment There is no specific treatment for the dermatitis other than soothing lotions and patience until healing proceeds.

Beneficial uses The roasted nuts with shells removed are a significant agricultural item, bringing a premium price on the world market. Each 100 grams of the mature seed contain about 540 calories, with 17% protein and 50% oil along with nutritionally significant amounts of vitamins and minerals. The light-yellow oil from the cashew kernel is composed of palmitic, stearic, oleic and linoleic fatty acids and is considered a specialty culinary item. An oil obtained by steam distillation of the leaves has tranquilizing effects similar to but less potent than those of chlorpromazine. Anacardic acid and cardol from CNSL have shown experimental activity against several human cancers. CNSL has been used in the formulation phenol-formaldehyde resins and plastics with good abrasive resisting and electrical-insulating properties. The resins are used in brake linings, clutch facings, acid-proof cement and tile, electrical insulators and insulating varnish. CNSL has also been used to protect wood carvings and other small craft items against termite and insect damage. The oil has been used by indigenous people to produce scars as social or tribal marks.

Cashew shells distilled under vacuum yield an oil which is the nontoxic phenol cardanol. Cardanol may be treated with formaldehyde to produce a resin useful in formulations of varnish and plastics. CSNL combined with formaldehyde and hydrochloric acid polymerizes to produce an ion exchange resin used to remove minerals and soften hard water. Hydrogenated cardanol is used as a fixative in perfumes and a component in lacquers. The moderately hard, brownish-pink, easy-to-work wood is very strong and attractive, but susceptible to dry-wood termites. It has been used in construction of boats, cabinetry, furniture and packing cases and in general carpentry. The clear gum which exudes from the injured stem is marketed as cashawa gum. The gum is composed of arabinose, galactose, rhamnose and xylose. It is used in pharmaceuticals and has been incorporated in the glue of book bindings as an effective deterrent to insects. An ink has been made from the sap. The young leaves are eaten as a vegetable. The "apples" may be eaten fresh, made into chutney, preserves, pies or ice cream, or fermented into wine or vinegar. In some countries, the wine is further distilled into a good brandy. 100 grams of the fruit contain 260 mg of vitamin C or about 5 times the minimum daily requirement. The fruit juice is astringent and has been recommended for uterine problems and dropsy. Anacardic acid and several compounds derived from it have been found to be bactericidal against *Staphylococcus aureus*, *Mycobacterium smegmatis* and *Bacillus subtilis*. The compounds are antifungal against *Trichophyton mentagrophytes* and *Saccharomyces cerevisae* and show nematicidal action. Consumption of an alcohol extract of the bark is reported to lower the blood-sugar level within 20 minutes, reaching a maximum after 90 minutes and still evident after 3 hours.

An extract of the bark utilized in cardiac surgery elevated the cardiac glycogen and prevented the ouabain arrhythmias caused by glycogen depletion. The bark and leaf extract also is reported to reduce blood pressure and hyperglycemia to normal levels. A water extract of the bark and leaves is astringent and has been used internally to treat dysentery and externally as a mouthwash to relieve toothache and sore gums. Folk medicine has used the leaf tea to treat diabetes, indigestion and diarrhea. Bark extracts have been used to treat asthma, colds, diabetes, diarrhea, dysentery, elephantiasis, fevers and sore throat. The CNSL and the sap have been used to treat leprosy and fungus diseases, remove warts and corns and by primitive peoples as an arrow poison. Ayurvedic medicine uses the fruit as an aphrodisiac, to treat dysentery, fevers and piles, and to eliminate intestinal worms. Chinese medicine uses the oil from the fruit to treat leprosy and worms. The pharmacopeias of many countries include parts of this plant for medicinal applications. The Dutch use the leaves, the Mexicans the fruit and gum. The fruit is used by the French, Portuguese, and Venezuelans. The root is purgative. The plant in flower is very attractive to bees.

Notes The genus is named from Greek after the appearance of the nut, *ana* ("like") and *kardium* ("heart"). The species epithet refers to the western (American) origin of the plant. The tree is known by many similar common names (*caju* in Spanish and Portuguese, *acajou* in French, *kasjoe* in Dutch) which relate to its original Indian name. The average nut weighs 7 g (0.25 oz), of which 2/3 is shell. The kernel contains about 50% edible oil along with sucrose, starch and cellulose.

Christmas-bush
Comocladia dodonaea

p. 8

Anacardiaceae

Toxic properties The resin found in the sap and as a residue on the surface of the leaves contains a urushiol similar to but more potent than that found in poison oak and poison ivy.

Symptoms The slightest contact with the plant or its sap produces burning, itching skin eruptions after a period of several hours in sensitive individuals. The swelling and irritation may take several days to fully develop. Due to the delayed onset of symptoms, unknowing people may accumulate enough toxin to produce a rather severe dermatitis before they are aware of any difficulties with the plant. Once absorbed, the toxin is reputed to be capable of moving through the body and causing eruptions at other than the point of contact. The worst of the symptoms subside within two weeks of exposure, but evidence of the rash may remain for much longer. The toxin seems to be quite stable and may be acquired from dead leaves and from clothing which has had contact with the plant. Some people are not sensitive to the plant and others become sensitive only after repeated exposure.

Treatment The best treatment is prevention. When walking in the woods, watch for the plant and walk around it. When clearing land of the plant or trimming a plant grown as an ornamental, wear cloth gloves and a long-sleeved shirt. Launder both after use, and for good measure, take a shower to remove any urushiol which may have inadvertently contacted the skin.

Beneficial uses The attractive, red-brown wood is used for carving small, decorative items and is harmless when finished. It is a hardy, attractive plant and is sold by nurseries as an ornamental. It is reported that the foliage has been used to prepare an invigorating bath, but this is one of those unfortunate instances where a different plant with the same common name has been confused by a person recording folk medicinal uses. The intentional exposure of one's entire body to this plant could result in a catastrophic case of dermatitis.

Notes The genus name is from the Greek *kome,* meaning "hair" or "tuft," and *klados,* meaning "branch," referring to the leaves crowded at the tips of the branches. This plant is also known as poison ash, cock's spur and chicharron.

Mango   p. 9
Mangifera indica   Anacardiaceae

Toxic properties The sap, leaves, green fruit and skin of ripe fruit contain the resin 3-pentadecyl catechol, which induces allergic reactions in some. There is a high degree of cross-reaction between those sensitive to poison ivy and those responding adversely to the resin of mango. The leaves contain mangiferine, a glucoside; mangiferol, a phenol; and the resinous mangiferic acid.

Symptoms The adverse reactions to the various parts of the plant are due to the development of an allergic sensitivity to the resin and not to the presence of anything inherently harmful. An individual may handle mangos for many years, then become sensitized by the scratch of a freshly broken twig or sap in the eye. Children may develop a dermatitis on the extremities from climbing the tree. The sap released from the stem when the fruit is picked is particularly troublesome and should be rinsed from the fruit. Contact with the sap, leaves and green fruit induces an itching rash, blisters and swelling. Itching and swelling around the eyes, nose and corners of the mouth are typical systemic responses of sensitive individuals to exposure to the sap or excessive consumption of the fruit. With no introduction or mention of symptoms, an experienced tropical physician once greeted the author with the comment, "What are you here to see me about other than the fact that you have been eating too many mangos?"

Systemic shock reaction is possible but extremely rare as a result of eating mango. Even sensitive individuals can consume a modest amount of the fruit if someone else removes the peel. The flowers give off a compound (other than pollen) which irritates the respiratory system and may even cause rash and swelling of the eyelids. Smoke from the burning leaves and branches causes respiratory distress and may cause dermatitis. Excessive consumption of the fruit may lead to diarrhea, hemorrhoids, kidney problems and other digestive system malfunctions, yet certain individuals may regularly eat green and ripe mangos, including the skin, for years with no adverse effects. Most cultivated varieties are less inclined to produce the allergic response than the unimproved wild types. The smoke from burning leaves or wood may cause a inflammation of the eyes, a skin rash and itching. Heavy consumption of unripe mangos has caused human kidney inflammation. Consumption of considerable amounts of mature mango leaves by cattle over a period of time can cause illness and mortality in livestock. The urine of affected individuals contains a bright-yellow salt of euxanthic acid, which is used as a dye.

Treatment Avoiding the skin of the fruit and its sap, along with washing hands and face after handling or consuming the fruit, reduces the allergic response. Commercial poison ivy resin removal products will remove the allergic resins from the skin.

Beneficial uses This is one of the most popular and widespread of all tropical fruits. World production of the fruit is about 15 million metric tons. Mangos are abundantly produced and consumed in the tropics and are now widely exported to colder climates. The green fruits, tart with citric acid, are frequently preserved as various jams and chutneys or made into hot or sweet pickles to be used as appetizers. The ripe fruits are probably best when eaten raw, but are also sun dried or made into sweet conserves, beverages or many other commercial products. The young leaves are high in vitamin C and are cooked and used as a vegetable in Java and the Philippines. Chinese medicine has used the smoke from smoldering leaves to treat hiccups and throat ailments and a poultice to treat asthma, coughs and skin problems. Folk medicine has used the leaf tea to treat asthma, bronchitis, coughs, diabetes, diarrhea, hypertension, insomnia and whooping cough.

The bark tea is used to treat asthma, bronchitis, diarrhea, gonorrhea and skin diseases. The bitter seed has been used to eliminate intestinal worms and to treat diarrhea and piles. The root bark is an astringent which reduces diarrhea. The roasted or boiled seeds may be eaten or ground into a nutritious flour. The hard, tough, strong wood is used for chopping blocks, construction carpentry, crates, cooperage and carts. A variety with streaked wood has been used for fine furniture. The bark and leaves yield a yellow dye called peori or Indian yellow. Livestock eat the fruit, but cattle sometimes choke to death when attempting to swallow an entire fruit. Bees visit the flowers to produce a delicious, aromatic, thick, amber honey.

Notes The genus name is from the Latin words meaning "mango bearing." The species name is from the Latin, referring to its origin in India. The composition of the fruit of different varieties of mangos varies considerably, but all contain abundant vitamin C when green and are composed primarily of sugars when ripe, with lesser amounts of organic acids, proteins, pectin and various volatile substances providing aroma. The leaves contain a large percentage of euxanthic acid as well as euxanthon, hippuric acid, benzoic acid and tannin. The resin in the fruit contains mangiferene, mangiferic acid and mangiferol. The bark contains up to 20% tannin and quercetin. The seed kernel is predominantly starch, but also contains tannin, gallic acid and an oil.

Poisonwood   p. 10
Metopium toxiferum   Anacardiaceae

Toxic properties Allergic contact dermatitis may occur due to urushiols similar to those of poison ivy and its kin. The toxic urushiol is a component of the resin found in canals that permeate the roots, stems and leaves of the plant. Contact with any part of the plant can produce allergic contact dermatitis in sensitized individuals. Urushiol is found in the green fruit, but seems to be absent in the ripe berries and flowers. Many birds and mammals consume the fruits with impunity. The toxin is a mixture of similar compounds, each with a catechol ring possessing a 15 carbon side chain with 1 or 2 double bonds. In the presence of oxygen in the skin, the urushiol is converted to a highly reactive quinone, which binds to the proteins in the skin. The toxin is very stable and remains in the leaves and the wood as long as they remain intact.

Symptoms Skin contact with the sap or foliage produces delayed development of an itching, painful, swollen rash and blisters. There is considerable variation in response, from seeming absolute immunity to extreme sensitivity in which the symptoms persist for more than 3 months after exposure. Symptoms may be delayed for several days after exposure. Exposure may be from direct skin contact with the foliage or sap of the plant or with tools or clothing which have been in contact with the plant. Exposure to the smoke from the burning plant has produced symptoms in many people due to the transport of small droplets of the urushiol. Woodworkers are irritated by the dust from the wood. The sap often leaves black oxidized marks on the skin.

Treatment Abundant washing with soap and water promptly after exposure will reduce or eliminate the dermatitis.

Beneficial uses Fruit pigeons and particularly white-crowned pigeons feed heavily on the fruit when it is ripe. The hard, heavy, brown wood streaked with red is harmless when finished and takes a fine polish. The resinous exudate from the bark has been used in folk medicine to prepare diuretics, emetics and purgatives. This plant produces a copious nectar yielding an excellent-quality amber honey.

Notes The generic name is from the Greek name for the similar gum of an African tree. The species name is Latin, meaning "poison carrying." The tree also is called poison tree.

Brazilian pepper   p. 10
Schinus terebinthifolius   Anacardiaceae

Toxic properties All parts of the tree contain a volatile irritant compound which produces an allergic reaction. The fruits contain the phenol cardanol, which has been shown experimentally to be an irritant (with a long latency period) of the skin and mucous membranes. The fruit also contains an essential oil composed of several monoterpenes, with phellandrene the most prevalent. It is likely that the phenols and terpenes have a synergistic effect in irritating the skin. The probable contribution to the fruit toxicity by the contained triterpenes terebinthone and schinol has not been confirmed. The volatile monoterpenes from the fruit and flowers may be responsible for the respiratory irritation often associated with this plant. The resin from the trunk is reported to contain cardol, and the symptoms it produces certainly are consistent with that conclusion. Fruit and leaves falling into ponds cause fish to die.

Symptoms When the tree is in bloom, it may produce sneezing, asthma, sinus congestion, eye irritation and headache in sensitive individuals. Although the symptoms are similar to those of hay fever, it is a volatile irritant chemical emanating from the blossoms that causes the problem and not the sticky pollen, which is seldom carried in the wind. Consumption of the berries fresh, or when dried as the spice "pink pepper," can result in irritation of the throat, vomiting and diarrhea, hemorrhoids, rash, swelling of the face or eyelids, shortness of breath, lassitude, shivering and violent headache. Some individuals may develop a severe contact dermatitis of the anus or hemorrhoids. Physical contact with the plant or its sap may produce an itchy rash over the entire body. The lesions may become quite severe in the case of individuals such as woodcutters or tree trimmers repeatedly exposed to the sap. Horses develop dermatitis from rubbing on the tree, and those eating the berries may develop a fatal colic. Cattle develop enteritis and other allergic symptoms from eating the foliage or berries, but goats seem to be able to consume modest amounts with impunity. Birds feeding heavily on the berries become intoxicated and unable to fly. Massive mortalities of wild birds have been attributed to feeding on *Schinus* berries.

Treatment The offending trees should be removed by nonsensitive people whenever possible. Use of the fruit as a spice should always be avoided. Therapy is usually not required except to treat dehydration and provide symptomatic relief of respiratory difficulties, digestive upset or anal itching.

Beneficial uses The compound pentagalloglucose, extracted from the leaves, was found in experiments to have antiviral effects and a potential for treating gout and kidney stones. The dried fruits of this tree and the similar *Schinus molle* are sometimes marketed (dangerously) as pink pepper at exaggerated prices. Although they are hot to the mouth, they are not true peppers and their use as a condiment is dangerous. The wood is used for rough construction and railroad ties.

Homeopathic medicine uses the plant for arthritis, diarrhea and gout. Folk medicine has used the bark tea as an astringent, a tonic and a stimulant. Externally, it has been applied to relieve gout, rheumatism, sores and syphilis. A resin from the trunk is marketed as *balsamo de missiones* in Spanish-speaking communities and has been used to treat tumors and rheumatism. The leaves and fruit have been used as folk medicinal remedies for arthritis, bronchitis, diarrhea, gout, rheumatism, sciatica, sexual inertia, syphilis, tumors and ulcers. The medium amber honey with a distinct peppery taste sells as a novelty item called "Florida holly honey." The majority of the harvest generally is considered suitable only for inclusion in baked goods and the manufacture of other confectionery items.

Notes The genus is the Greek name for the mastic tree, which has a similar resinous juice. The species name is a hybrid of Greek *teribinthos* and Latin *folius* and refers to the resemblance of the leaves to those of the genus *Terebinthus*. Other common names for this plant include Christmas berry, holly and aroeira. Beekeepers are among the few people who look on this tree with favor because when it blooms, it produces an abundant nectar. The resinous sap turns black upon exposure to air, as in many of the other Anacardiaceae.

Poison ivy, oak, sumac
Toxicodendron radicans, toxicarium, vernix   p. 11
                                                         Anacardiaceae

Toxic properties The toxic component is an urushiol which is a component of the clear resin found in canals that permeate the leaves and the phloem of the roots and stems of the plant. The resin oxidizes and polymerizes to a black, gummy substance within a few hours of its exposure to air. Urushiol is found in the green fruit, but is absent in the ripe berries and flowers. The toxin is a catechol ring with an unbranched 15 to 17 carbon side chain with up to 3 unsaturated bonds. The urushiol of each species of *Toxicodendron* has a predominant characteristic structure with lesser, variable amounts of related compounds. *Toxicodendron radicans* has mostly 15 carbon chains with 2 double bonds. *Toxicodendron toxicarium* has mostly 17 carbon chains with 3 double bonds and *Toxicodendron vernix* has only 15 carbon chains, most with a double bond. In the presence of oxygen in the skin, the urushiol is converted to a highly reactive quinone which binds to the proteins in the skin. The toxin is very stable and remains in the leaves and the phloem of the plant as long as they remain intact.

Symptoms The typical symptom is the delayed, gradual development of an itching rash leading to blisters. Symptoms may be delayed for several days after exposure. Exposure may be from direct skin contact with the foliage or resin of the plant or tools, clothing or animals which have been in contact with the plant. Tales abound of methods of accidental exposure such as hanging a towel on a bush while swimming, then applying the toxin thoroughly to the body when drying off. Dogs romping in the woods may bring home the toxin on their coat. Droplets of the resin carried in the smoke from the burning plant have produced symptoms in many people. About 50% of the adult population of the U.S. is sensitive to poison ivy. Whites are more likely to be acutely sensitive than are blacks. In all groups, the degree of sensitivity to the toxin and the response to treatments varies greatly among individuals. Very sensitive individuals may show severe systemic manifestations, including fatal kidney degeneration or other terminal complications.

Treatment The best treatment continues to be avoidance. Knowing the appearance of the plant (3 leaflets per leaf) and avoiding contact is best. Prompt removal of the toxin is essential to prevent the chemical reactions which bind it to the skin. Once the symptoms have manifested themselves, toxin removal efforts are of no avail. If you have knowingly been exposed, washing the exposed skin with soap and water within 30 minutes of exposure will help prevent symptoms in all but the most sensitive individuals. Solvents such as acetone and alcohol have been found effective in removing the toxin from the skin. Washing with soap and water within two hours of exposure reduces the severity of reaction. Topical application of barrier creams, composed of activated clay or various proprietary formulations, prior to exposure can reduce the amount of urushiol contacting and being absorbed by the skin. Treatment by oxypropyleneamine salts of linoleic acid and various proprietary compounds may prevent dermatitis after exposure by aiding in the removal of the toxin.

Treatment of fully developed acute poison ivy dermatitis traditionally has been with bland ointments until the affliction runs its natural course. Topical application of fluorinated corticosteroids prior to the appearance of blisters may significantly reduce evolution and spread of the dermatitis, but they should be used cautiously to avoid systemic effects. Various systemic corticosteroids and adrenocorticotropic hormone can dramatically reduce severe acute dermatitis. None of the postsymptomatic treatments significantly abbreviate the course of the affliction. Folk medicine has developed a wide array of treatments for the dermatitis, ranging from the mildly ameliorative to the bizarre and preposterous. Although investigators have been looking for a solution, no safe, effective and reliable method has been widely marketed yet that can reduce the systemic sensitivity of individuals to these attractive but irritating plants.

Beneficial uses The wood is not poisonous, but is seldom used except for novelty items due to its small size. Flickers and other birds and mammals use the fruit as a significant part of their winter diet. Many small creatures use poison ivy thickets as a refuge. Horses, cows, goats, sheep, deer and many other herbivores browse the foliage with no adverse effects. The Dutch use the vigorous low-growth habit of the plant to help stabilize dikes. The fruit clusters are sometimes employed in dried flower arrangements. Native Americans have used the sap to dye baskets and as an ingredient in an arrow poison. The stems have been used in basket making and the leaves have been used in baking acorn meal bread. Folk medicine has used the sap to eliminate warts and ringworm and to treat rattlesnake bite. Bees harvest the nectar to produce an excellent nontoxic honey. The honey does not seem to produce allergic symptoms, but has a bitter flavor to some individuals.

Notes The generic name is from the Greek *toxicon*, "poison," and *dendron*, "tree." The species epithet is the Latin word meaning "rooting stem." The toxic compounds were named urushiols by the Japanese chemists who discovered them because they were obtained from *kiurushi*, the Japanese name for the sap. *T. toxicarium* is also known as *T. diversilobium*. The wood from this genus is not poisonous because the toxic resin is carried in ducts beneath the bark. The unsubstantiated myth that Native Americans have been rendered immune to the dermatitis caused by poison ivy due to the regular consumption of its leaves is dangerous folklore. Ingestion of leaves or fruit may bring about acute inflammation of the membranes contacted and kidney failure due to the absorbed poison.

Yellow allamanda p. 13
Allamanda cathartica Apocynaceae

Toxic properties The leaves and sap or tea produced from the plant may produce a severe, persistent diarrhea. The agent in the sap causing skin irritation has not been identified.

Symptoms The sap causes an itching rash on the skin and burns the eyes. It produces nausea, elevated temperature and local irritation of the mouth and lips when small amounts are consumed. Consumption of larger amounts leads to vomiting and diarrhea.

Treatment The sap should be immediately washed from the skin and the eyes should be abundantly rinsed with clean water. The diarrhea is usually self-limiting, but in severe cases may require fluid and electrolyte replacement.

Beneficial uses This plant is a very popular ornamental used in planters, as a border plant and draped from retaining walls. It makes a good hedge, but the required trimming interferes with its flowering. The root serves as a good stock for grafting of other less vigorous species. The stem and root contain two rare lactones, plumericin and isoplumericin, which are active against polio virus and pathogenic fungi. The two compounds show promise as marine antifouling agents, as they are both highly toxic to barnacles and algae. These compounds previously were known only from frangipani. More recently allamandin, a new iridoid lactone, has been found effective against leukemia in experimental mice and against human nasopharyngeal cancer in cell culture. Folk medicine has used the leaf tea as a laxative and emetic. The root tea has been used in various formulations to treat malarial symptoms. The sap was used as a purgative and to eliminate intestinal worms, but the practice has been recognized as unsafe and mostly abandoned.

Notes The genus is named to honor F. Allamand, a Swiss botanist who collected seeds in Surinam and sent them to Linnaeus in 1770. The species is from the Latin word meaning "cathartic." Other common names include yellow trumpet and campanilla. The plant has a triterpene ursolic acid and an iridoid glycoside named plumieride in the leaves. The flowers have yielded kaempferol and quercetin.

Periwinkle   p. 14
Catharanthus roseus   Apocynaceae

Toxic properties There have been more than 70 indole alkaloids found in this plant, along with several glycosides. Most of them are pharmacologically active, with varying degrees of toxicity.

Symptoms The plant is toxic to livestock and the leaves produce vomiting in humans. The root bark is purgative and its toxic effects have been used to eliminate intestinal worms. When the side effects of therapeutic use (euphoria and hallucinations) became known to the public, a transient epidemic of smoking the dried leaves occurred. The epidemic ended as the users discovered the many additional adverse effects on the kidneys, liver and nervous system.

Treatment If consumed, the leaves or flowers should be removed from the stomach as soon as possible to prevent additional absorption of alkaloids.

Beneficial uses Modern medicine has found the alkaloids from this species to be the most important contribution from the plant kingdom to cancer therapy. Vinblastine sulfate, sold as Velban, and Vincristine sulfate, sold as Oncovin, used separately and in combination with other drugs, have resulted in 99% remission in acute lymphocytic anemia, 80% remission in Wilms tumor and Hodgkin's disease (people cured of the latter disease sometimes completely lost interest in sex), 70% remission in choriocarcinoma and 50% remission in Burkitts lymphoma. The alkaloid ajmalicine has been broadly applied in circulatory diseases, particularly in the obstruction of normal blood flow to the brain. Acylindolic acid from the roots and leaves has been found experimentally to be active against influenza virus. Considering the variety of alkaloids and other compounds present in this plant, it is likely that many other legitimate medical uses will be discovered.

Folk medicine has used the flowers, leaves and roots variously in the treatment of diabetes, dysentery, hypertension, malaria and rheumatism and to regulate menstruation. The flower tea has been used to treat asthma and flatulence. The root tea is astringent and is used to treat dysentery, fevers and upset stomach and to induce abortion. The plant tea is also used to treat diabetes, tuberculosis and high blood pressure and to regulate menstruation. As recently as 1988, a semiscientific report with no data was published on the curative properties for diabetics of daily consumption of an extract of the vegetative parts of *Catharanthus*. As with many of the reports of this nature, no preparation details or dosage amounts were revealed.

Notes The generic name is from the Greek *katharos,* meaning "pure," and *anthos,* meaning "flower." The species name is from the Latin word meaning "rosy" or "pale pink," in reference to the most common color of the flowers. This plant has also been known as *Vinca rosea* and *Lochnera rosea*. It has contributed more to saving lives of cancer victims than any other plant, yet the therapeutic potential of this plant was once entirely overlooked by the National Cancer Institute screening program for plants with activity against cancer.

Oleander p. 14
Nerium oleander Apocynaceae

Toxic properties The plant is poisonous due to the presence of several cardiac glucosides which the body selectively concentrates in the heart muscle. The flowers, seeds, leaves, latex, bark, roots, smoke from burning wood, and even honey prepared from the nectar are all poisonous. Humans and livestock have been poisoned by consuming water into which leaves and flowers have fallen. A baby is reported to have died after consuming the milk from a cow which consumed oleander foliage. The principal toxic chemicals are a group of cardiac glycosides similar in action to digitalis. The two dominant glycosides are neriin (nerioside) and oleandrin (oleandroside), with lesser amounts of folinerin, rosagenin, cornevin, pseudocuranine, rutin and cortenerin. Total glycosides of each plant part make up as much as 0.5% of the weight of the part at various seasons of the year. Red-flowered forms tend to have more glycosides than white-flowered forms. There is considerable variation of the specific influence of each glycoside structure on heart function.

Symptoms Humans have been made ill by chewing a single flower or leaf or accidentally transferring a minute amount of the sap to the mouth after handling the cut plant or flowers. Children have died from sucking the nectar or eating several of the flowers, and adults have been made ill or died after using the handy straight twigs to skewer meat for cooking. Smoke from the burning wood is toxic, as is the water in which flowers have been sustained. Symptoms associated with oleander consumption include decreased, sometimes irregular pulse rate, abdominal cramps with vomiting and diarrhea (sometimes bloody), dilated pupils, dizziness, persistent headache, fatigue, drowsiness and loss of visual acuity, with blurred or aberrant color vision. With more serious cases, convulsions, respiratory failure, coma and death are possible. A red flush surrounding the mouth has been reported in intoxications and fatalities.

Skin contact with the leaves and sap produces a rash in some people. The presence of oleander flowers in a closed room has affected some individuals. Livestock consuming the leaves show a diminished cardiac reserve with rapidly increasing heart rate with small effort. They also show vomiting, diarrhea, stupor, trembling, convulsions and paralysis. Before death, animals become restless and uneasy, with excitement leading to sudden death. Horses and cows have been killed by the consumption of 20 g (0.7 oz) of fresh leaves and sheep by 1/10 of that amount. Many horses and cows have been killed by the accidental or nefarious introduction of oleander leaves into their feed.

Treatment The stomach contents should be removed immediately and should be replaced later by a slurry of activated charcoal. Rapid-acting cathartics should be given to remove any plant remains from the intestinal tract. Once absorbed from the intestinal tract, the oleandrin and nerin are only slowly eliminated from the body. Monitoring of serum potassium and cardiac function by electrocardiograph is suggested in all but the most trivial cases. The TDx Digoxin II serum assay has been found to accurately report the effective levels of toxin in the blood serum. Treatment with atropine and propranolol was the treatment of choice for many years. Modern technology has developed a new treatment with an intravenous solution of digoxin-specific Fab antibody fragments, marketed as Digibind, which is reported to bring about rapid improvement in human cardiac function by chemically binding to the toxin. Experimental treatment of dogs poisoned with oleander leaves has confirmed the efficacy of digoxin-specific antibodies in relieving the symptoms. A poison-control center should be contacted for the most up-to-date treatment protocols.

Beneficial uses The leaves contain a latex which has been considered as a commercial source for rubber. Extracts of the plant have been used as an insecticide. Purified oleandrin is in the Russian pharmacopoeia and has been substituted at the dose of 0.2 mg/day for digitalis and ouabain in treating cardiac insufficiency. It may be tolerated when the latter two are not and it also regularizes cardiac flutter and fibrillation. Like the digitalis glucosides, it may take several weeks to eliminate 50% of the amount in the blood from the body. The toxic diuretic neriin has only 1/40 of the toxicity of ouabain. Folinerin has 2 or 3 times the cardiac action of digitoxin, but is less cumulative and more easily removed by the kidneys. It is being investigated for possible clinical use. Cornerin produces diuresis and has been used successfully in heart therapy. It is less cumulative than the digitalis glucoside. The antibiotic oleandomycin extracted from the plant is effective against penicillin-resistant *Staphylococcus*. The seeds have been used to poison pigs and jackals. The bark powder is used as a rodenticide and an insecticide.

Externally, folk medicine has used the plant to treat snakebite, scabies, mange and fleas. The leaves and bark macerated in coconut oil are particularly effective in eliminating external parasites on humans and domestic animals. Traditional Chinese medicine has used the plant in various formulations to treat heart problems. In spite of the dangers, folk medicine has used this plant internally as a purge and vermifuge. It is probably very effective in both internal roles if the patient survives the treatment. Although folk medicine has used this plant internally in numerous other ways, they will not be chronicled here due to their potential danger. Various preparations from the plant have been used as arrow poisons.

Notes The genus is the original Greek name for oleander used by Dioscorides. The species name makes reference to the olive (*Olea*)-like leaves. Other common names include rose laurel or rose bay. The nectar which bees gather from the flowers yields a toxic honey. The history of difficulties with this plant extends back to Alexander's army, which lost horses which fed on the foliage and lost soldiers who used skewers of the wood to grill meat. The ancient Greeks considered oleander poisonous to all four-footed beasts. This plant contains a veritable cornucopia of toxic compounds far too numerous and complex to be discussed here.

Bitter ash p. 15
Rauvolfia nitida Apocynaceae

Toxic properties The major indole alkaloid present in the plant is rauwolscine, which strongly lowers the blood pressure without sedative effects. Deserpidine, employed in modern medicine as a tranquilizer and to lower blood pressure, is extracted from the roots. The root contains reserpine, one of the early tranquilizers. The roots and other parts of the plant contain at least 25 other pharmaceutically active alkaloids which have not been fully investigated. The ripe fruit is the most common source of human poisonings.

Symptoms The sap may cause skin irritation and blisters on contact. Consumption of the fruit produces constriction of the throat and an intense burning sensation, with extreme thirst, vomiting, diarrhea and convulsions. The central nervous system is most affected. The extremities often feel cold as the heart rate and blood pressure are depressed, while mental processes are slowed and calmed. Deaths have occurred due to overdose of folk medicines and the individual alkaloids. Strange, intense dreams may occur under the influence of the alkaloids, and severe depression may lead to suicide. Chronic exposure may lead to hormone imbalance resulting in lack of menstruation or reduced sexuality in males.

Treatment The sap should be washed from the skin as soon as possible. The stomach contents should be removed by vomiting or gastric lavage if the fruit has been eaten.

Beneficial uses The drug reserpine and related alkaloids from the roots of the plants of this genus are widely used in modern medicine as tranquilizers and sedatives and to lower blood pressure by vasodilation. Administered over several days, reserpine reduces the arterial blood pressure and slows the heart while producing a relaxed tranquil state with reduced responsiveness to external stimuli. Indian holy men have used the drug to promote tranquil meditative states, while Ayurvedic medicine and herbalists in primitive societies have used the drug for centuries to calm the ravings of the mentally disturbed and as a sedative. The drug was used for several centuries in Europe to treat anxiety before it was introduced to clinical medicine in the U.S. in 1948. Modern psychiatry has used the drug effectively to treat various mental disturbances and to aid withdrawal from opiate addiction. Deserpidine (also called canascine or recanescine), used as a tranquilizer and to lower blood pressure, is extracted commercially from the roots. Yohimbine slows the heart, lowers blood pressure and induces a mentally hypnotic state.

This plant is a good source of many other pharmaceutically active alkaloids and commercial pharmaceuticals which can be extracted at less cost than chemical synthesis in the laboratory. The drugs extracted from this genus have served as a model for the chemical synthesis of other frequently used drugs such as chlorpromazine, Milton and Equanil. The hard, lightweight, clear, yellow wood has been used for furniture, carvings and musical instruments. The juice of the fruit has been used as an ink or a dye. Folk medicine has used various decoctions of the plant to treat fever, gum disease, malaria, skin diseases, snakebite, sore throat, syphilis and many other ailments. Use of the mashed fruit to cure various external parasites, including mange on dogs, may be effective due to the toxic components of the fruit being lethal to the vermin.

Notes The genus honors Leonhard Rauwolf, a German physician who died in 1596. The species name is from the Latin word meaning "shining," in reference to the shiny leaves. This plant has also been known as *Rauvolfia tetraphylla* (a shrub) and *Rauvolfia heterophylla*. It has also been called bitterbush, pinque-pinque and milkbush.

Crape jasmine   p. 16
Tabernaemontana divaricata   Apocynaceae

Toxic properties A number of indole alkaloids have been reported from the leaves, bark and roots. Coronaridine is the principal cytotoxic alkaloid. The structure of those named apparicine, mehranine, lahoricine, stapfinine and ervatinine has been described in the technical literature. Nine other indole alkaloids have been identified in extracts of the plant. Also present are 6 different triterpenoids and a sterol.

Symptoms Various questionable sources report that the seeds produce a delirium similar to that of *Datura*. Other sources report alkaloids which are violent cardiac poisons and others with paralytic properties. Mortalities have resulted from the consumption of a tea made from this plant.

Treatment As the specific toxins have not been identified, removal of the plant remains from the stomach, then administration of a slurry of activated charcoal, would be a conservative approach.

Beneficial uses The primary use of this plant is as an ornamental shrub. The wood is burned as a component of incense and extracted for use in perfumery. The pulp surrounding the seeds has been used as a red dye. Crude extracts have shown anticancer activity and the plant has been used to treat cancer in Taiwan. Some fractions of methanol extracts have shown strong blood pressure-lowering effects and others have shown uterus-relaxing properties in the presence of oxytocin. The compound coronaridine, extracted from the roots, has been found to prevent pregnancy in rats at an oral dose of 5 mg/kg per day. Both may provide significant medical benefits when the responsible compounds are purified and identified. Folk medicine has used this plant internally for the treatment of coughs, diarrhea, epilepsy, intestinal worms, ophthalmia and scorpion sting, and externally on wounds and inflammations. The pulped root has been used to treat sore eyes, skin ulcers and ulcers of the nose caused by tertiary syphilis.

Notes The generic name, which honors J. T. von Bergzabern (d. 1590), who Latinized his name as Tabernaemontanus, is derived from an early common name for the plant. The species epithet is Latin, meaning "spreading" or "straggly." This plant has also been known as *Nerium divaricatum, Ervatamia divericata, Tabernaemontana coronaria,* and *Ervatamia coronaria.* This generic name is a Latinization of Maylaylan (a language of the west coast of India), meaning "flower of Nandi" (a common temple offering). The fruits are used as an ingredient in arrow poison. Chemical analysis of the leaves has revealed coronaridine, hydroxyindolenine, voacristine, isovoacristine, voacangine, vaocristine, tabernaemontanine, dregamine, sitosterol and others.

Luckynut p. 16
Thevetia peruviana   Apocynaceae

Toxic properties All parts of the plant are poisonous due to the presence of at least 8 cardiac glucosides which the body selectively concentrates in the heart muscle. The seed kernels contain the highest concentration of toxins. Thevetin A and B are found in the seed kernels, leaves and the bark of the roots and stems. Thevetin is a bitter glucoside with a potent cardiac action similar to that of digitalis, with 1/8 the strength of ouabain. Thevetin has a stimulant effect on the smooth muscles of the intestine, bladder, uterus and blood-vessel walls. Favorable but not invariable results have been reported from patients treated with thevetin for congestive heart failure. The margin between therapeutic and toxic dose is probably too small to allow thevetin's regular therapeutic use. Several reports in the literature are in strong disagreement on the chemical formula for thevetin.

Thevetoxin is another cardiotoxic glucoside that is less toxic than thevetin and also stimulates smooth muscles of the intestine, uterus, bronchus, blood vessels and heart. Neriifolin (kokilphin), peruvoside and ruvoside are also found in the kernel and have actions similar to those of digitalis. The digestive tract symptoms may be caused by a different toxin.

Symptoms Consumption of a seed or leaves causes a dry or numb burning in the mouth and throat, followed by abdominal pain, vomiting, diarrhea, dilated pupils, dizziness, slow, irregular heartbeat with high blood pressure and sometimes death by heart failure. Symptoms may be delayed as much as 24 hours after consumption of the toxic plant part. A red flush surrounding the mouth has been reported in intoxications and fatalities. One seed or 2 leaves have been lethal to children and 8 seeds have produced death in an adult. Lethal human poisonings by accident, folk medicine overdose, suicide and homicide have been reported regularly. The symptoms progress rapidly to the extent that death may ensue within 2 hours. The sap causes skin irritation and sometimes blistering. A dosage of 15 g (0.5 oz) of the leaves may prove fatal to horses or cattle.

Treatment The contents of the stomach should be promptly removed. The patient may then be given activated charcoal. Cardiac function should be monitored by electrocardiograph, and serum potassium should be frequently examined. The adverse effect of the toxins on heart function may last as long as 5 days. The TDx Digoxin II serum assay has been found to accurately report the effective levels of toxin in the blood serum. Supportive and symptomatic treatment should be provided as needed. Use of the newly developed digitalis-specific antibodies may be appropriate in severe, life-threatening intoxications.

Beneficial uses The tree is most commonly cultivated for its attractive yellow flowers. The lightweight, hard, gray wood is easily worked and has a fine texture. A bright-yellow, nontoxic oil suitable for food use or soap making can be extracted from the seeds. The oil (nontoxic when pure) is composed primarily of oleic, linoleic, stearic and palmitic acids. The flesh of the fruit covering the seed is reputed to be edible. The folk medicinal use of the seed oil in treating burns and infected wounds has been supported by the discovery that one of the fractions distilled from the seed oil is active against the common infective bacteria *Staphylococcus aureus, Streptococcus pyrogenes, Escherichia coli* and *Pseudomonas aeruginosa*. The seeds are used as beads on necklaces and carried as pocket charms. The pulp of the fruit is reported to be eaten with impunity by chickens, livestock and humans, but this would seem imprudent. Thevetin has been used medically to treat mild myocardial insufficiency in the presence of digitalis intolerance. In Russia, it is used for cardiac insufficiency with shortness of breath, and for ventricular insufficiency due to high blood pressure and atherosclerosis.

With partial hydrolysis and the loss of two glucose units, Thevetin A yields the therapeutic cardioactive drug peruvoside. One author extolling the virtues of peruvoside outlines the advantages over ouabain in treating congestive heart failure as: 1) its origin from a native plant makes the supply reliable and inexpensive, 2) an oral dose is completely absorbed from the intestinal tract, 3) it has a rapid action (4 to 6 hours), 4) it has a wide safety margin and 5) it is less cumulative and has no side effects. One research report states that when Thevetin B is stripped of its sugar component, it is identical to digitoxin (a clinically useful cardiac glycoside). The presence of the anticancer compounds cerberin and ursolic acid may be the basis for the use of leaf poultices to treat tumors in Latin America. Folk medicine has used the sap to treat aching teeth, chronic sores, ulcers and mange. The bark, leaves, roots and seeds, although often recognized as toxic, have been used in various formulations to treat bladder stones, edema, fevers, insomnia, hemorrhoids,

146

malaria and snakebite and to intoxicate fish for capture. Juice extracted from the leaves has been mixed with a meat bait and used to kill nuisance tigers near Malay villages. Aucubine, an iridoid heteroside extract from the leaves and fruit, is an effective insecticide.

Notes The generic epithet honors Andre Thevet (1502–1592), a French monk who traveled in Brazil. The species name means "of Peruvian origin." This plant is also commonly known as *Thevetia neriifolia* and *Cerbera thevetia*. Other common names include yellow oleander, lucky-seed, cathartic bark, campanilla, campañero and French trumpet-flower. The seeds have been imported to the U.S. as a source for a heart-regulating medication. The oil from the seeds is composed (in descending order) of oleic, palmitic, linoleic and stearic fatty acids. The latex has been used to poison arrows.

Dumb cane
Dieffenbachia seguine

p. 17
Araceae

Toxic properties Sharp needles of calcium oxalate (raphides) are fired from a chambered envelope in special cells when the plant tissue is subjected to mechanical pressure such as chewing. These specialized cells sequentially fire their entire contents of several hundred needles when activated. The juice from the leaves contains a proteolytic enzyme called dumbcain which produces itching, swelling and pain when introduced to the victims' tissues by the many calcium oxalate needle punctures. It has also been theorized that the needles injure mast cells, releasing large amounts of histamine into the injured tissue. Other proteolytic enzymes such as those in the stinging hairs of *Mucuna pruriens* and those in scorpion venom are known to induce these symptoms via the production of kinins. A cyanogenic glycoside has been reported from *Dieffenbachia seguine, D. picta and D. amoena,* but it seems unlikely to contribute significantly to the toxicity.

Symptoms Exposure of the skin to crushed stems, leaves or juice may produce a transient itching, burning and local inflammation. Chewing the stems or leaves of *Dieffenbachia* results in rapid development of profuse salivation, burning pain, redness and swelling of the tongue and throat. Loss of the ability to speak (accounting for the common name of the plant) due to inflammation of the larynx may occur and last for several days. Swelling of the mouth and throat may be so severe as to induce fatal choking. Microscopic examination of the tissues shows swelling, vascular congestion and degeneration of the basement membrane. Free oxalic acid introduced to the tissues by the raphides may contribute to the symptoms, but the total amount of oxalate consumed is rarely enough to produce systemic effects. The worst of the pain and swelling usually subsides within less than a week, but lesions may take several weeks to heal. Exposure of the eyes to the juice results in intense pain, aversion to light, swelling of the eyelids and corneal abrasions. Close examination of the cornea with magnification frequently shows lesions and the needle-like calcium oxalate crystals. The free-flowing sap without the needles produces a keratoconjuctivitis.

Full recovery usually takes place within 3 to 4 weeks, but corneal opacity may persist in severe cases. The sap also produces sterility in rats. In experiments with rabbits, there seems to be considerable variation among species in the potency of the juice in causing eye irritation, with declining toxicity from *D. seguine* to *D. picta* to *D. amoena*. Other reports show the stem to be more potent than the leaf and old growth more toxic than new growth.

Treatment The swelling in the mouth and throat recede slowly after about 4 days and resolve in 12 days without treatment, but some relief has been reported by quickly applying lime juice to dissolve the calcium oxalate needles which have been discharged and or have penetrated the tissues. The severe pain does not regress significantly for about 8 days. Experimentally, glycerin has been found to inhibit discharge of the needles. Diphenhydramine has been recommended to reduce the swelling. Analgesics such as xylocaine or meperidine, demulcents and cool beverages held in the mouth may bring partial relief. Antihistamines may provide relief, but usually do not alter the course of recovery. Tracheotomy has been necessary in severe cases. The calcium oxalate is not present in sufficient amount to cause systematic toxicity. The 3- to 4-week period required for healing of eye injuries produced by the juice has been experimentally halved in animals with the application of 1% ethylmorphine (to produce improved permeability of the cornea) and 2% disodium edetate (for dissolution of the calcium oxalate needles).

Beneficial uses The literature reports the belief that chewing a leaf induces male sterility for 24 hours. The use is reported in populations as diverse as officers of the Third Reich during World War II and South American natives. Rural people on several West Indian islands continue to claim the leaf is an effective male contraceptive. Perhaps the leaves have a lower concentration of the harmful raphides than the stem. Ingestion or injection of the juice has been reported to cause sterility in lab animals after several weeks. Homeopathic medicine has used the alcoholic extract from this plant to treat frigidity and impotence. Folk medicine has used the plant to treat cancer, corns, coma, dropsy, frigidity, gout, impotence, warts and yaws. The plant is used as an insecticide. A food-grade starch can be extracted from the stem, and the root can be boiled and used as food. The leaf juice produces a permanent dark-brown color when applied to cloth and has been used as a laundry mark.

Notes The genus is named for the early 19th-century Austrian botanist and gardener J. F. Dieffenbach (1790–1864). This plant has also been called *Arum seguinum* and *Caladium seguinum*. South American natives reportedly use the juice in arrow poison and to sterilize their enemies. Slaves have been punished by rubbing the cut stems of this plant in their mouths. A criminal has used the stem to render a witness mute, resulting in acquittal. The stems have been used to cause the crystallization of sugar in boiling vats, perhaps due to the raphides' providing a nucleus of crystallization.

Pothos   p. 18
Epipremnum aureum   Araceae

Toxic properties The toxic component has not been identified, but it is suspected that the primary irritant is calcium oxalate needles.

Symptoms The clear, watery sap strongly irritates the skin and eyes. Internally, it irritates the mouth and throat and may produce diarrhea. An allergic contact dermatitis has been reported as a result of contact with the leaves.

Treatment The internal symptoms are usually self-limiting, but dehydration may require fluid replacement in young children. Skin irritation may be relieved by topical application of corticosteroid creams.

Beneficial uses This plant is widely used as a component of tropical landscaping and as an indoor plant climbing from pots and hanging from baskets. The air roots have been woven into strong and durable baskets. The young shoots have been fed to horses to rid them of intestinal worms.

Notes The genus is from the Greek *epi,* meaning "upon" and *premnon,* meaning "trunk," from its habit of climbing tree trunks. The species epithet is from the Latin word meaning "golden," referring to the color of the flowering stalk. This plant is the victim of considerable taxonomic shuffling. It has appeared in the older literature as *Scindapsus aureus, Pothos aureus* and *Rhaphidophora aurea*. Its common names also include hunter's robe, taro vine and trepapalo amarillo. The spathe or the fruits have been used in arrow poisons. It is unlikely that these components themselves are toxic; rather, the hundreds of sharp calcium oxalate needles contribute considerably to the absorption of additional poison by the afflicted tissues.

Monstera

p. 18

Monstera deliciosa

Araceae

Toxic properties Cells containing calcium oxalate needles have been identified in the aerial roots and other parts of the plant.

Symptoms Irritation of the mouth and throat results from biting the leaves or consumption of the sap or less-than-ripe fruit. Juice from the damaged plant is severely irritating to the eyes.

Treatment The pain and swelling begin to decline within 2 to 4 days and are usually gone within 2 weeks. In severe cases, swelling of the tongue and throat may require a tracheotomy to allow respiration. The dissolving of the calcium oxalate needles can be accelerated by the application of lime juice or vinegar. The oxalate content is not sufficient to produce systemic poisoning.

Beneficial uses The fully ripe fruit has a pineapple-like flavor and is eaten fresh or pulped and included in beverages. Early jungle explorers found the ripe fruit to be a "delicious rescue amidst the green hell."

Notes The generic name is from the monstrous appearance of the leaves. The species name refers to the flavor of the fruit. The plant has also been known as *Philodendron pertusum*. This plant is also known by the common names ceriman, Swiss cheese plant, cut leaf philodendron and Mexican breadfruit.

Aralia

p. 19

Polyscias guilfoylei

Araliaceae

Toxic properties The plant contains saponins, triterpenic glycosides and other unidentified irritating agents.

Symptoms Repeated contact with the plant may produce inflammation and a local or widespread rash. Systemic allergic reaction with swelling is reported.

Treatment Pyrabenzamine hydrochloride is reported to have provided significant relief to an individual with running sores from exposure to this plant.

Beneficial uses This plant is widely used as an ornamental. The saponins in the leaves have considerable toxicity to the snails which serve as a host of schistosomiasis. The leaf tea is used by folk medicine to treat colds, headache and stones in the urinary tract.

Notes The generic name is Greek *poly*, "many," and *skias*, "umbel," because the main umbel is divided into many lesser umbels. The species honors W. R. Guilfoyle, a 19th-century collector of plants from Polynesia. This plant is also known as *Aralia guilfoylei*. It is also known by the common name of coffee tree. Several cultivars with varying foliage are available from nurseries.

Fishtail palm
Caryota mitis

p. 20
Arecaceae

Toxic properties Sharply pointed calcium oxalate crystals in the pulp of the mature fruit are externally irritating and internally toxic. The fibrous hairs at the base of the leaf stalks produce a skin irritation. The fruit, leaf and stem all contain various alkaloids.

Symptoms Skin punctures by the crystalline needles from the juice produce an intense itching within seconds, followed by redness and swelling which may last 8 to 12 hours. The juice produces an intensely uncomfortable eye irritation.

Treatment The needles may be removed from the skin by application of adhesive tape. Otherwise, pain and swelling recede slowly without treatment.

Beneficial uses The kernels of the *Caryota* fruit and the terminal bud are edible. Muslims have used the seeds as beads and others have used them as buttons. A single tapped inflorescence can produce as much as 14 liters of sap per day, which may be consumed fresh as neera, fermented to produce toddy, then distilled to produce the more potent arrack. Sugar (called jaggery), with 2.3% protein and significant vitamin content, is refined from the sap of the tapped inflorescence. A strong, elastic fiber from the leaf sheath commercially marketed as kitool (or kitul) is used to make brooms, brushes, bowstrings, fish line, nets, ropes and hats. The stronger and more coarse fiber from the petioles is traditionally used to make ropes to tether wild elephants. A woolly fiber found on the underside of leaves is used to caulk boats and to cauterize wounds. The trunk may be split and used for house rafters, water conduits, rice pounders or drums. The core of the growing tip may be cooked and eaten as hearts of palm and is a favorite food of elephants. The flowers and expectant buds are used to promote the growth of hair. Starch called bastard sago, extracted from the pith of *Caryota,* is a significant part of the diet in several less-developed societies. This plant is used as an attractive ornamental in landscaping. It sometimes is grown in large pots as an indoor plant in colder climates.

Notes The genus is from the Greek *karyon,* "nut." The species epithet is from the Latin word meaning "harmless" or "mild," referring to the lack of spines. When these palms reach maturity, they begin flowering at the bases of the upper leaves. The flower stalks emerge sequentially lower on the trunk until they reach the lowest leaves, at which point the tree dies. The mashed fruit has been put into wells to annoy people. Bathing with this water causes an intense itching of the skin and an acute inflammation of the eyes. One particularly virulent poisoning recipe calls for mixing the stinging pulp of the fruit with the hairs from bamboo and extract of toad.

Butterfly weed
Asclepias curassavica

Toxic properties The cardiotoxic glycosides calotropin, calactin, calotoxin and many similar chemical compounds are present in all parts of the plant. Various cardenolides make up about 50% of the latex of the plant. Consumption of one pound of leaves per hundred pounds of body weight usually is fatal to livestock. The plant is not particularly palatable to livestock due to the presence of the polyphenols quercetin and kaempferol, but will be consumed if other fodder is scarce. Sheep are the most common victims, but cattle and horses may also succumb when pastures are overgrazed or in dry times when other pasture species have been consumed.

Symptoms The latex causes dermatitis in some people. In humans, the symptoms after consumption of the latex or a tea made from the plant include vomiting, diarrhea, salivation, fever, dilation of pupils, labored breathing and coronary insufficiency. Afflicted animals become dull and inattentive within several hours. They may stagger and lose muscular control before falling to the ground. The pulse will be found to be weak and accelerated and the breathing labored. Death may ensue rapidly when large amounts of the weed have been consumed.

Treatment Any plant residue should be removed from the stomach if vomiting has not already accomplished this. Ingestion of significant amounts by humans should be treated as a possible digitalis overdose, with electrocardiograph assessment and supportive therapy as needed.

Beneficial uses This plant is cultivated as an ornamental for its vividly colorful flowers. Attempts have been made to use the plant latex as a source of rubber and the plant as a source of fuel and hydrocarbons. Although these industrial uses are all possible, they are not economically viable ventures. The silky hairs from the seeds have been used to stuff pillows. Calotropin from this plant has been found to be active against human cancer of the nasal passages. Other unsubstantiated reports claim it is effective against cancers of the stomach, intestines, uterus, and kidneys. Experimentally, calotropin has the same potency as ouabain on heart function. Folk medicine has used the tea from this plant to induce abortions and to treat cancer, fevers, skin diseases, tumors, venereal disease and warts. Tea from the roots has been used as an emetic or tonic and to treat gonorrhea, intestinal worms, pneumonia, vaginal infections and various cancers of the digestive and reproductive tracts, and externally to reduce malignant tumors. The powdered root is a strong emetic and has been used as an adulterant and substitute for *Ipecacuanha*. The latex has been used to treat and disintegrate aching teeth, as a remedy for poisonous bites and to remove warts. Bees produce a light, high-quality honey from the nectar.

Notes The genus is named for Asklepios, the Greek god of medicine. The species name refers to the island of Curacao. This plant has also been known as *Asclepias syriaca*. It has also been known by the common names Kittie McWanie, swallow-wort, red-headed cotton, snake weed, horse killer, rat killer and wild ipecac. The latex contains nitrogen- and sulfur-containing cardenolides with thiasoline/thiazolidine rings. The compounds include voruscharin, uscharidin and calotropagenin. The leaf contains the cardenolides uscharidin, uscharin, calotropin, calactin, calotropagenin and calotoxin. The root contains the cardenolide aglycones azarigenin, corotoxigenin, corotoxigenin and coroglaucigenin. Caterpillars of monarch butterflies retain the toxins to the extent that birds eating the adults immediately vomit and subsequently avoid other individuals of the species.

Giant milkweed
Calotropis procera

p. 21
Asclepiadaceae

Toxic properties All parts of the plant contain several similar cardenolides, the most prevalent of which is the cardiac glycoside calotropin, which in low doses slows and strengthens the heartbeat. An overdose stops the heart. Also present are the glycosides uscharin, uscharidin, calotoxin, voruscharin and calactin, with action on the heart similar to that of ouabain. The latex is an irritant to the skin and mucous membranes, as it contains calcium oxalate needles and a proteolytic enzyme variously identified as trypsin or calotropain which is more active than papain or bromelin. Some individuals develop an acute allergic dermatitis after contact with the plant or its latex. The honey produced from the flowers of this plant is reported to be poisonous.

Symptoms Consumption of any part of the plant or a tea made from it may produce digitalis-like symptoms. Experimental inclusion of the leaves in the diet of sheep at the rate of 0.5% to 1% of body weight per day resulted in diarrhea, breathing difficulties, loss of hair and declining condition, then delayed death with liver, kidney, heart and lung deterioration. Experimental inclusion of 0.017% of the body weight of latex in the diet of goats produced hyperexitability, breathing difficulties, rapid heartbeat and convulsions resulting in death. Postmortem examination showed adverse effects to the intestines, liver, kidneys, spleen, heart and lungs. The blood of animals consuming the latex was dark-brown and clotted slowly. The latex is particularly noxious in the eye, and causes pain and swelling of mucous membranes. It produces irritation and blistering of the skin. The skin irritation may be an allergic response rather than direct irritation. Internally, it is irritating to mucous membranes and is a strong purgative.

Treatment Ingestion of significant amounts should be treated as a digitalis overdose, with electrocardiograph assessment. Atropine has been used to treat the irritation of sap in the eye. Externally, offending sap should be rinsed away with water and the irritated skin treated with soothing ointments.

Beneficial uses Calotropin has been found by modern medicine to have antitumor activity. An unidentified compound in the latex has been found to have anticoagulant and fibrinolytic properties. Calotropin has been listed in the U.S. Pharmacopoeia as a cardiotonic drug. Calotropain, the proteolytic enzyme present in the latex, is effective in removing intestinal worms and has been used in the tanning industry to remove the hair from hides. The flowers are strung into rosaries. The silky, fluffy hairs from the seeds have been spun into thread and used as a stuffing for pillows and dolls. The white, soft wood is pithy and seldom used except as fuel. A scheme has been considered to grow the plant commercially in northern Australia as a feedstock for the production of wood alcohol. As a byproduct of this industry, it has been proposed that the high protein content of the leaves could be utilized as a supplemental feed for livestock. Contrary to other reports, the experimental inclusion of small amounts of fresh leaves from northern Australia in livestock rations produced no adverse pathological or clinical symptoms.

The plant has been considered for the photosynthetic production of liquid fuels and has been found to contain an extractable hydrocarbon in the leaves with a heat value similar to that of fuel oil or gasoline. When methanol and hexane are used for extraction, the residual leaf material is essentially free of cardenolides and has a high protein content suitable as a supplemental livestock feed. A fine fiber is harvested from the inner bark and used for spinning and weaving. The stem yields a strong fiber which exceeds cotton in tensile strength and is used to make bowstrings, fishing lines, harnesses, nets, and ropes. It is reported that sheep and goats can eat the flowers without harm. The flowers have been preserved in sugar and used as a confection in Java. An alcoholic extract of the flowers reduced experimental inflammation and fever as effectively as aspirin. The alcoholic extract also inhibited the release of prostaglandins, which in themselves contribute to swelling and inflammation. The flower extract also showed antibacterial activity against gram-positive and gram-negative bacteria. A chloroform extract of the root bark showed reduction of inflammation and fever. The leaves have been used for their insecticidal properties. The "juice" has been used to curdle milk for cheese and as a component in fermented beer.

Folk medicine has used the leaves externally to treat colds, headaches, rheumatism and swollen or sprained feet. The dried leaves are smoked to relieve asthma. The powdered flowers have been used to treat colds, coughs, asthma and indigestion. The sap has been inserted in aching tooth cavities, and has been used to induce abortions and to treat elephantiasis, leprosy, syphilis and intestinal worms. A twig inserted into the cervix of a pregnant woman is said to induce abortion within a day or two. The latex has been used as an effective component of arrow poisons.

Notes The genus is from the Greek *kalos*, "beautiful," and *tropis*, "ship" or "keel." The species epithet is the Latin word meaning "tall." Other common names are mudar (in India), calotrope, wild down and French cotton. In India, the plant was used in folk medicine as a substitute for mercury to treat syphilis and thus became known as vegetable mercury. The flowers are sacred in India and are involved with the gods Kama and Siva. The latex has been used in homicide and suicide and as a cattle poison. Egyptian men have used the sap to produce eye damage to avoid military conscription. It has been used to poison arrows and spears of primitive peoples on both sides of the Atlantic (probably effectively, due to its influence on the heart). Monarch butterflies use the plant in all stages of development and are able to store the toxins as a defense against predation.

Purple allamanda
Cryptostegia grandiflora

p. 22
Asclepiadaceae

Toxic properties The cardiac glycosides cryptograndoside A and B, with actions similar to digitalis, are found primarily in the leaves, but are also present in the bark. The cardioactive components seem to be absent from the latex. The latex is irritating to the skin. A dust given off by the dried foliage is extremely irritating to the respiratory system.

Symptoms Tea from the bark or leaves has caused lethal effects in humans by homicide, suicide and accidental medicinal overdose. Survivors of poisoning by a tea made from the bark have shown a persistent low blood sugar for 5 weeks or more. The effect has been experimentally produced in rabbits. A few drops of the sap taken internally produce vomiting and diarrhea. The sap is injurious to the eye and causes serious skin irritation. Cattle and sheep are prone to graze on the regrowth of the plant after it has been burned or cut back. After feeding in the foliage for several days, they are prone to sudden death after brief exercise. The symptoms at death and postmortem examination are consistent with induced high blood pressure and subsequent cardiac arrest. Horses are particularly sensitive to the toxin in this plant and have died after accidental inclusion of small amounts of the dried leaves in their feed. Dust from the dried foliage causes violent coughing, swelling and irritation of the nasal membranes and may blister the eyelids.

Treatment Sap in the eye should be rinsed copiously with clean water immediately. For internal consumption, electrocardiograph performance and serum potassium levels should be monitored after the stomach is emptied.

Treatment for digoxin poisoning with atropine for conduction defects and phenytoin for rhythm disturbances may be required.

Beneficial uses The human nasal cancer-inhibiting steroids gitoxigenin, 16-anhydrogitoxigen, 16-propionylgitoxigenin and oleandrigenin 3-rhamnoside have been extracted from this plant. The sap produces permanent marks on fabrics. Folk medicine has used the sap to treat athlete's foot, eczema and other skin diseases. It has also been used to remove calluses and warts. The vine was planted in the 19th century as a source of rubber from the abundant latex, but harvest is not economically viable in current times. A high-quality fiber can be derived from the stem, but with too much difficulty to make it commercially worthwhile.

Notes The generic name is derived from the Greek *kryptos,* meaning "hidden," and *stego,* meaning "to cover," referring to the scales in the flower throat which cover the anthers. The species epithet is Latin, meaning "large flowered." Other common names for this plant are pink allamanda and rubber vine.

Calabash
Crescentia cujete

p. 23
Bignoniaceae

Toxic properties The raw pulp of the fruit is emetic and purgative, is reputed to be toxic to birds and induces abortion in cattle. Hydrocyanic acid has been reported in the pulp of the fruit. Other sources report that the pulp is eaten in times of scarcity, and that chickens fed the cooked pulp produce large eggs.

Symptoms It is suspected of causing abortion in cattle. A widely consumed folk medicinal syrup made from the cooked fruit pulp and experimentally added to the drinking water of mice resulted in numerous cases of a leukemia-type lymphoma.

Treatment Consumption of products made from the fruit pulp should be avoided.

Beneficial uses The tough rind of the fruits is dried or smoked and used for bowls, cups, jugs and other containers called by the generic term calabash. The dried rind also is used as a sounding board or other component of wind, string and percussion instruments of great variety. The wood is hard, strong and flexible, leading to its use in tool handles, ox yokes, wagon parts, stirrups and saddle frames. The bark seems to provide a particularly salubrious foundation for epiphytic plants in the wild and is used by horticulturists as a foundation for growing orchids and bromeliads. The seeds may be cooked and eaten. Folk medicine has used the astringent leaf tea to treat colds, diarrhea, dysentery, and headaches, to make the hair grow, as a tonic, and as a vaginal douche. The raw or cooked fruit pulp is used to treat asthma, coral cuts, dermatitis, diabetes, diarrhea, fevers, spider bites, sprains, tuberculosis and tumors, to eliminate fleas on domestic animals, and as a preventive and treatment for manchineel dermatitis. The roasted fruit or its juice are said to stimulate menstruation, induce childbirth and promote the passing of a dead fetus. The powdered, roasted seeds are used internally and externally to treat the bites of poisonous snakes.

Notes The genus is named in honor of Pietro de Crescenzi (1230–1321), an Italian agricultural author. The species is derived from an original Brazilian name. This plant has also been known as *Crescentia cuneiflora*. The common name is an Anglicization of similar names in French, Spanish and, interestingly, Arabic (*qar'ah yabisah*). Many varied shapes of calabash may be obtained by tying and training the developing fruit. Up to 37% of the weight of the seed may be extracted as an edible oil similar to peanut oil or olive oil. The oil contains 59% oleic acid, 19% linoleic acid and 1.6% linoleic acid, with the remainder being saturated acids.

Wild cinnamon

p. 24

Canella winterana

Canellaceae

Toxic properties On distillation, the bark yields a volatile oil having an odor similar to that of a mixture of cajeput and cloves. The oil contains benzoyleugenol, caryophyllene resins, canellal, cineol, clovanidiol, helicid, mannitol, myristicin, l-pinene, numerous drimane sesquiterpenes, warburganal and an unidentified bitter component. The bark is insecticidal and has been macerated into a slurry for use as a fish poison. The stems and leaves are toxic to chickens.

Symptoms The insecticidal and fish-toxic nature of the bark suggest it may have an action similar to rotenone, which is extracted from another plant. An infusion of the leaves in rum produces a pleasant, spicy beverage which leaves a fierce hangover. The leaves and stems have been shown to be toxic to poultry in feeding trials.

Treatment Consumption in moderation as a spice leaves no known adverse effects, but the considerable number of potentially harmful compounds indicates that restraint should be exercised in chronic use.

Beneficial uses The bark was formerly an article of commerce in the pharmaceutical trade in both the U.S. and Europe. A sesquiterpene dialdehyde called canellal or muzigadial, isolated from bark, is antifungal, antimicrobial and cytotoxic and inhibits insect feeding. The bark and leaves are still used locally in the West Indies as a mild, aromatic bitter to spice beverages and season food. The bark is incorporated in certain proprietary tonics, spice blends and aromatic smoking tobaccos. The powdered bark is listed as a drug in the British Pharmaceutical Codex and mixed with aloe is marketed as hiera picra. The dried green fruits are sold locally and used as a spice. The volatile oils extracted from the leaves have been used as ingredients in blending perfumes.

The strong, hard, heavy, blackish wood takes a smooth finish and a high polish. It is used as beams, plow frames, poles, posts and in small decorative articles. In folk medicine, the inner bark tea is taken to treat fevers, relieve indigestion and induce abortion and is gargled to treat inflamed tonsils. Externally, the leaves are applied to relieve rheumatism and headaches. A rum extract of the bark is used as a liniment to treat rheumatism and other pains. The liquor may be consumed to treat stomach pains. The finely chipped wood may be smoked alone or with other plant materials to relieve headache and hangover.

Notes The generic name is derived from Latin *canna*, "a reed," alluding to the rolled bark of this tree as sometimes seen in commerce. The generic name is also reputed to originate from the Latin word for cinnamon as it is used in Spanish *(canela)* and French *(canelle)*. The species epithet honors a Captain Winter who first introduced the bark of the tree to Europe, but other authors state that it is derived from the winter flowering habit of the tree. This tree is also known as *Canella alba* and *Winterana canella*. Other common names are barbasco, caneel and canella.

Australian beefwood
Casuarina equisetifolia

p. 25
Casuarinaceae

Toxic properties The pollen produces allergic reaction in some individuals. The roots secrete allelopathic toxins which prevent other plants from growing nearby.

Symptoms In sensitized individuals, sneezing, runny nose, inflamed and itching eyes, cough and asthma result from exposure to the pollen.

Treatment It is best to stay upwind when trees are in flower. Remaining inside an air-conditioned house seems to relieve most of the symptoms in a neighborhood with many flowering trees.

Beneficial uses The reddish-brown, fine-textured, hard, heavy heartwood is used in rough construction as beams and rafters, but must be treated before use in permanent structures because it is susceptible to attack by dry-wood termites. This tree often has been used to stabilize coastal sand dunes and to provide lumber and fuel in an otherwise inhospitable environment. The tree responds well to hedging and has been used in topiary work. The bark has 18% tannin and has been used for tanning. Casuarin, the coloring matter in the bark, has been used for dyeing. The astringent bark is used in folk medicine to treat beriberi, diarrhea and dysentery.

Notes The genus is named for the resemblance of the drooping branches and foliage to the feathers of the cassowary (*Casuarinus*). The species is named after the superficial resemblance of the leaves to those of the plant *Equisetum*. This tree is known by many other common names, including Australian pine (although it is not a pine; it only resembles one), casuarina, she-oak, ironwood and horsetail. The fruits have been found to contain ellagic acid, beta-sitosterol and trifolin.

Purple queen
Tradescantia pallida

p. 26
Commelinaceae

Toxic properties The irritants have not been identified, but it is speculated that oxalate crystals, along with a caustic chemical, are present.

Symptoms Persons sensitive to this plant show a redness, burning and itching of the skin, sometimes followed by blisters after exposure to the sap. Sap in the eye causes great discomfort and inflammation.

Treatment Washing with soap and water followed by a soothing ointment greatly alleviate the irritation.

Beneficial uses It is a hardy, aggressive plant frequently used as a ground cover. The anthocyanins which give the plant its color are of interest as food colorants because they are stable in beverages and retain their color at a pH above 4, unlike the anthocyanins of cranberries, which are vivid at a pH of 1 but almost colorless at 4. The anthocyanins are also about 29 times more stable than other similar vegetable colors in nonsugared drinks.

Notes The genus is named for J. Tradescant (d. 1638), a British gardener. The species epithet is a Latin word meaning "pale." This plant has also been known as *Setcreasa pallida* and *Setcreasa purpurea*. Another common name is purple heart. Many other cultivars and species of this genus, such as *Tradescantia zebrina* (Wandering jew) also known as *Zebrina pendula*, with similar medical characteristics, are grown as ornamentals in both the house and garden.

Oyster lily
Tradescantia spathacea

p. 26
Commelinaceae

Toxic properties The watery sap has an unidentified irritating chemical in it.

Symptoms The watery juice reddens the skin and provokes a stinging, itching, burning sensation. Gardeners often suffer a rash from extensive skin contact with the plant or its sap. Transient pain and irritation result from a droplet of the juice contacting the eye. Burning of the mouth and throat, stomach pain and intestinal irritation result from internal use.

Treatment The best treatment is to remove the offending sap from the skin as soon as possible with soap and water, or with clean water if the sap is in the eye.

Beneficial uses Folk medicine has used the astringent juice of the leaves or a decoction to stop bleeding, both externally and internally. The leaves are used to induce abortions in the first few months of pregnancy. Various formulations of the plant have used to treat coughs, pulmonary complaints and venereal diseases and to promote healing. The reddening effect has led to the sap's being utilized cosmetically as a rouge.

Notes The genus is named for J. Tradescant (d. 1638), a British gardener. The species name is a Latin word referring to the spathe-like large bracts which envelope the flower. This plant is also known as *Rhoeo discolor* and more frequently in the horticulture trade as *Rhoeo spathaceae*. Other common names include boat lily, Moses-in-a-boat and sangria. Wildlife, including deer, raccoons and ducks, eat the plant.

Balsam apple
Momordica charantia

p. 28
Cucurbitaceae

Toxic properties The leaves, seeds and unripe fruit contain the bitter, cathartic, alkaloid momordicine and the toxalbumin momordin, which inhibits protein synthesis in the intestinal wall. The seeds contain a purgative oil with cucurbitacins. The ripe fruit contains the steroid glucoside charantin, which has a hypoglycemic action.

Symptoms The seeds and body of the ripe fruit produce severe vomiting and diarrhea and are also reported to induce abortion. Human deaths have occurred after consuming the fruit. Pigs that feed regularly on the leaves and fruits are reputed to have unpalatable meat.

Treatment The influence of the intoxication is usually self-limiting, but fluid and electrolyte replacement may be necessary. Intestinal function may be sufficiently compromised to require alternative nutritive sources.

Beneficial uses Experimentally, the roots have shown antibiotic activity. The ripe fruit extract lowers blood glucose concentrations independently of intestinal absorption of glucose in mice. The two different compounds seem to have an effect outside the pancreas. In China, the dried, powdered fruit is reported to alleviate the high levels of glucose in the blood and urine of moderate diabetics by increasing carbohydrate metabolism. Medical reports show an increased glucose tolerance without a serum insulin increase when the fruit juice is included in the diet. A polypeptide isolated from the fruit and seeds is called p-insulin. It has 17 different amino acids and a molecular weight of about 11,000 and shares most of the amino acids with bovine insulin, but does not cross-react with bovine insulin.

Subcutaneous administration of p-insulin to diabetic gerbils, monkeys and humans produced a reduction in blood sugar after 4 to 8 hours. Cataract is delayed and blood sugar is reduced in diabetic rats fed an aqueous extract of the fruit. The fruit extract included in the diet of male dogs brought about sterility without altering general metabolic activities. In gerbils, a fruit extract disrupted spermatogenisis without affecting the seminal vesicle or prostate. The roots cause abortion, perhaps due to the presence of charantin, serotonin, diosgenin and beta-sitosterol, all of which may stimulate the uterus. Charantin causes abortion in rabbits and inhibits fetal development in rats. Proteins called momorcharins, extracted from the seeds, have been found to have abortifacient, immunosuppressive and antitumor activities. An extract from the fruit has experimentally killed human leukemic lymphocytes while not affecting the viability of normal lymphocytes.

The fruit has been used in folk medicine to treat asthma, bacterial dysentery, cholera, colitis, diabetes, gonorrhea, gout, high blood pressure, leprosy, malaria, psoriasis, rheumatism, scabies, snakebite, ulcers and other afflictions. The marginally ripe fruit has been soaked in whiskey with rock sugar to produce a product used as cough syrup. The juice of the ripe fruit has been used to treat worms, hemorrhoids, gout, leprosy and rheumatism, along with liver and spleen ailments. An extract from the root has been used to induce abortions, promote passage of bladder stones, treat hemorrhoids and malaria and as a component of aphrodisiac concoctions. The vine and leaves produce a very bitter tonic which is used to treat colds, colitis, constipation, fevers, high blood pressure, influenza, kidney stones, malaria and intestinal worms. It is widely used to induce abortion. An aqueous extract of the plant has been found in experiments to be effective in eliminating the parasitic worm *Haemonchus contortus* in goats. The poulticed leaves are used to treat the sore eyes of elephants.

The arils of the ripe wild fruit may be eaten fresh and the young leaves may be used as a vegetable after boiling in two changes of water to reduce the momordicine content responsible for the bitterness. There are many cultivated forms of this plant which bear fruit in different seasons, of different colors, shapes and sizes (to 30 cm [12 in] long)and different culinary characteristics. The young fruits of cultivated forms may be cooked as a seasoning with poultry, fish or meat. They may also be dried or pickled in brine with or without spices. The bitterness of mature fruits can be reduced by boiling, salting or other methods of extracting the juice; then they are used in curries, stews, chop suey or stuffing for meat dishes. The fruit is a good source of iron, phosphorus and vitamin C. The leaves are a good source of B vitamins. Bees seem to prefer the nectar and orange pollen of this plant over most other species.

Notes The genus is from the Latin word meaning "to bite," in reference to the seeds of some species which have a jagged edge as if bitten. This plant has also been called *Momordica cylindrica*. Other frequently used common names are maiden apple, bitter gourd, bitter melon, balsam pear, wild cucumber, bitter cucumber, cerasee, karela and cundeamor. This and the very similar *Momordica balsamina* (more slender, with smaller leaves and fruits) are frequently confused and the name balsam apple is applied to each. A Japanese patent (76 07.111) has been issued for deriving an insulin-like compound from *Momordica charantia*.

Coontie   p. 29
Zamia pumila   Cycadaceae

Toxic properties The toxic glucoside cycasin is found ubiquitously and exclusively in the 10 genera of living cycads. The toxin is present in the root, seeds and foliage. Cycasin loses a glucose molecule in the digestive system and becomes the physiologically active methylazoxymethanol upon absorption from the intestinal tract. The toxin must be removed from the root or seeds by grating and washing before human consumption is safe. Wash-water from preparation of the root may contain enough toxin to be lethal to livestock. Acute doses of cycasin in humans and animals produces severe liver damage. Chronic exposure of rats, mice and guinea pigs produces tumors at numerous sites in the body. Other experiments have shown cycasin to be a very strong mutagen. Long-term, chronic, low-level exposure results in permanent central nervous system damage. The neurotoxic compound beta-N-methyl-alpha-B-diaminoproprionic acid has been isolated from the seeds and is believed to be the cause of neurological symptoms in man. Azoxyglycosides have been identified in the pollen.

Symptoms In humans and in livestock, the consumption of the inadequately detoxified roots or seeds results in two syndromes. One is a short-onset digestive upset with continuous severe vomiting, which may be followed by diarrhea. The more severe syndrome is depression with a gradually develop-ing nonreversible neurological paralysis. In humans, neurological symptoms similar to those of amyotrophic lateral sclerosis gradually develop. As the life expectancy of native peoples has increased, the prevalence of symptoms in their populations has increased due to the greater time available for symp-toms to develop. In cattle, consumption of the foliage produces a disease called wobbles. The nonreversible condition results in a peculiar stance and uncoordinated stumbling gait in which they tend to throw their feet lateral-ly. Sheep have also been lost due to their grazing on cycad leaves and or seeds. The wind-dispersed pollen from male cycad cones causes coughing, sneezing and general irritation of the throat and respiratory passages.

Treatment The only known treatments are fluid and electrolyte replacement as needed and other symptomatic care. The allergic respiratory distress caused by the pollen may be avoided by removal of the male cones before they mature.

Beneficial uses The grated and washed roots contain up to 38% starch and 6% protein and are used to make the starchy soup "sofkee," which was a sta-ple of the traditional Seminole Indian diet and is a component of the diet of many other indigenous peoples. The extracted starch of the root has been marketed as arrowroot and used as the primary ingredient in biscuits, baby food, chocolates and spaghetti. The seeds of a similar species are reported to be eatable after grinding, washing and thorough cooking. Folk medicine has used the fruit as an ingredient in a therapeutic shampoo. The gum from the stem has been used to treat ulcers of the skin. The roots have been chewed to treat coughs and to improve the singing voice.

Notes The generic name is from a Latin word meaning "pine nut," in refer-ence to the form of the cones. The species epithet is from the Latin word meaning "dwarfed." This species is also known as *Z. floridana* and *Z. integri-folia*. Other common names are Seminole bread, Florida arrowroot, bay rush, sato (Jamaica), guayiga (Dominican Republic) and malunguey (Puerto Rico). One of Hernando DeSoto's men was fatally poisoned by the inadequately pre-pared root of *Zamia*, and during the Civil War, Union soldiers died in Florida when they ate bread made from *Zamia* roots.

Air potato   p. 29
Dioscorea bulbifera   Dioscoreaceae

Toxic properties The underground tuber contains the alkaloid dioscorine, the diterpene lactone diosbulbine, the steroidal sapogenin diosgenin, a poi-sonous glucoside and a series of complex phenolics. The aerial tuber contains alkaloids, saponins and oxalates, which decrease in abundance with maturity. The bitter characteristics are related to the abundance of numerous furanoid diterpenes.

Symptoms Consumption of the untreated aerial tuber results in a rapid stinging, burning and swelling of the mouth and tongue. Consumption of the underground tuber has been fatal to pigs.

Treatment Due to considerable variation in cultivars, local advice on preparation should be gathered before eating any of the tubers from this plant. In cases of suspected poisoning, stomach contents should be removed as soon as possible and supportive therapy provided as needed.

Beneficial uses The underground tuber may be eaten after elaborate preparation, including peeling, slicing, washing and boiling in water made alkaline by ashes or lime. The aerial tuber, weighing up to 2 kg (5 lb), may be similarly treated, but varieties in which the cut surface rapidly darkens are considered too toxic for consumption. The Asian varieties are generally superior in flavor and less toxic than those from Africa. Experiments with rats fed extracts of the tubers have supported the folk belief that consumption of the tubers before a meal reduces the appetite and subsequent food consumption. The root and the tuber have been found to have antibiotic effects on *Candida albicans*. The pharmacologically active compound allantoin has been found in the tubers. The mashed tubers have been used as a fish poison. Traditional Chinese medicine has used the plant to treat food poisoning, goiter, hernia and purulent inflammation. Folk medicine has used the steamed leaf to treat pinkeye and various parts of the plant to treat boils, diarrhea, hemorrhoids and syphilis.

Notes The genus is named for Pedanios Dioscorides, a 1st-century Greek herbalist who wrote a book which served as a foundation for botany after the Dark Ages. The species is a Latin word meaning "bearing bulbs," in reference to the aerial tubers. The yellow pigment is composed primarily of xanthophylls and not the nutritionally useful beta-carotene present in many root vegetables.

Tung tree   p. 30
Aleurites fordii   Euphorbiaceae

Toxic properties The attractive, pleasant-tasting seeds of both species are poisonous due to the presence of phorbol esters and an as-yet-unidentified toxic protein. The leaves are poisonous to livestock due to a toxic saponin. The seed meal remaining after oil extraction contains 2 toxins which are destroyed by cooking with an acid pH. Tung oil has produced itching, inflammation and blistering of the skin in sensitive individuals. The phorbol esters in the seed oil have been found to be cocarcinogenic.

Symptoms Chewing of the kernel and then spitting it out results in an irritation of the mouth and lips which makes eating or swallowing difficult for several hours. Consumption of a single tung nut or candlenut causes a feeling of warmth and abdominal pain within 30 minutes, followed by intense thirst and vomiting of bile. Within minutes or sometimes after as long as an hour, diarrhea develops to the extent that hospitalization is sometimes required. The pupils may be dilated but reactive, with other reflexes diminished or absent. Human poisonings have occurred as the result of substitution of tung oil for other edible oils such as rapeseed oil. In one such case in Germany, 190 people experienced severe vomiting and diarrhea after a baker used tung oil as an ingredient in pancakes. In severe cases, dehydration, weak rapid pulse and respiration or respiratory depression may lead to shock and death.

Cattle grazing on the foliage or eating the trimmings from the tree show loss of appetite, profuse watery-bloody diarrhea, weak listlessness, unthrifty appearance, emaciation and death. Symptoms in cattle may be delayed 3 to 7 days and death may occur as long as 3 weeks after consumption of the leaves. Postmortem examination shows an inflamed digestive tract, with enlarged liver and congested kidney and spleen. Pigs feeding on the seeds suffer liver and kidney damage along with inflammation of the gastrointestinal tract.

Treatment The stomach contents should be removed if natural processes have not already accomplished this. Oral magnesium sulfate has been recommended as reducing the severity of the symptoms. Electrolytes and fluids along with analgesics for gastrointestinal distress are often needed. Symptoms usually resolve themselves within 24 hours of rehydration.

Beneficial uses The drying oil pressed from the ground seeds of the tung nut is about 80% eleostearic acid and is used in making a water-resistant boat varnish, in the paint industry and in traditional oil-based putty. It produces a deep, mellow oil finish on fine hardwoods and dries at twice the rate of linseed oil. It is completely nontoxic when dried and can be used to finish wooden kitchen items such as bowls and spoons and cutting boards. The oil is burned for illumination and to produce the lamp black used in India ink. In China, it is used for waterproofing wood, cloth, paper, masonry, umbrellas, bamboo netting and as a motor fuel. Mixed with lime and hemp fibers, it makes an excellent caulking for wooden ships. Industrially, the oil is used to make adhesives, artificial leather, brake linings, electrical insulation, steam gaskets and about 800 additional patented applications. Fruit extracts have been found to be bactericidal.

Folk medicine has used the oil to treat boils, burns, high blood pressure, insanity, metal poisoning, ulcers and swellings. Traditional Chinese medicine has used the green fruit cooked with pork to treat anemia and ammenorrhea; and the oil to treat burns, edema, masturbation, scabies and traumatic bleeding from wounds. When applied to surgical wounds, the oil is reputed to suppress infection and inflammation, leaving no scar tissue. The methyl ester of alpha eleostearic acid has proven to be a very effective feeding deterrent for boll weevils. The wood of both species is soft and white and seldom used except as fuel. The candlenut is eaten after detoxification by roasting.

Notes The generic name is from the Greek word meaning "mealy" or "floury," referring to the appearance of the young leaves, twigs and fruits covered with abundant, silvery, star-shaped hairs. This tree is also known as the China wood oil tree. *Aleurites moluccana* is called kukui or candlenut after the habit of native peoples of burning the oily nuts as a source of illumination. The oil from candlenut contains no eleostearic acid and thus has little value in the varnish industry. The high incidence of nasopharyngeal cancer in China has been linked to chronic exposure to the tumor-promoting phorbol ester in *Aleurites fordii* growing as a common roadside tree. The tree was introduced to the U.S. when a consular officer in China sent seeds to the U.S.D.A. in 1904. In 1913, the first tung oil was extracted from the fruits of American-grown trees and by 1938, 200,000 acres were planted to tung trees. The U.S. consumes over 30 million pounds of tung oil per year. The tree is rather fussy in its temperature requirements: it needs 350 hours with the temperature below 45° in the winter, but it is killed by a freeze that drops below 10° for an appreciable time. It is also badly damaged if exposed to frost while in an active growth phase.

Maran   p. 31
Croton astroites   Euphorbiaceae

Toxic properties The toxin is similar or identical to the powerfully purgative croton oil. Croton oil is obtained from the seeds of *Croton tiglium* and several other species of *Croton*. It has been used as a purgative, but the use has generally been abandoned due to the violence of its action. The purgative component is soluble in oil and not soluble in alcohol, but can be absorbed through the skin. The skin-irritant component is soluble in alcohol. The oil induces skin tumors. The toxic, cocarcinogenic and inflammatory properties are produced by different chemical compounds which have not been definitely identified.

Symptoms A few drops of the oil produce intense stomach and intestinal contractions resulting in diarrhea. Externally, the oil is an irritant and may produce blistering of the skin.

Treatment The symptoms are usually self-limiting. In severe cases, rehydration therapy may be needed.

Beneficial uses A tea from the new growth is used in folk medicine to treat bladder trouble and gonorrhea.

Notes The generic name is Greek for "tick," based on the appearance of the seeds. The species epithet is from the Greek, meaning "star-like," referring to the star-like hairs on the leaves.

Broombush   p. 32
Croton betulinus   Euphorbiaceae

Toxic properties The toxic components have not been identified, but are presumed to be very similar to the other *Crotons*.

Symptoms A few drops of the oil produce intense stomach and intestinal contractions resulting in diarrhea. Externally, the oil is an irritant and may produce blistering of the skin.

Treatment The symptoms are usually self-limiting. In severe cases, rehydration therapy may be needed.

Beneficial uses The woody stems have been used in various handicrafts.

Notes The generic name is Greek for "tick." The species epithet is in reference to the leaf and stems having a superficial resemblance to the birch *(Betula)* tree.

White maran   p. 33
Croton discolor   Euphorbiaceae

Toxic properties The alkaloids crotonosine, 8,14-dihydrosalutaridine, methylcrotonosine, linearisine and discolorine have been isolated from the plant and contribute to its toxicity.

Symptoms A few drops of the oil from the seeds or the sap which oozes from broken petioles produce intense stomach and intestinal cramps and diarrhea. Externally, the oil is an irritant and may produce blistering of the skin.

Treatment The symptoms are usually self-limiting. In severe cases, rehydration therapy may be needed.

Beneficial uses Folk medicine has used a tea from the stewed leaves to treat rheumatism.

Notes The generic name is Greek for "tick." The species name is Latin and refers to the top and bottom of the leaves being two different colors.

Wild poinsettia   p. 33
Euphorbia cyathophora   Euphorbiaceae

Toxic properties The acrid latex contains neither alkaloids nor glucosides, but probably obtains its toxic properties due to a resin.

Symptoms The latex produces vomiting and diarrhea. In severe cases, dehydration and delirium may lead to death. Consumption of the leaf and latex have caused the death of a child.

Treatment The stomach should be emptied of its contents. Therapy for dehydration should be provided as needed.

Beneficial uses The latex is used in folk medicine as an antidote for irritation caused by other species of *Euphorbia*. The latex from the plant yields rubber. Extracts of the leaf and flower show activity against *Mycobacterium tuberculosis*. The abundant nectar is gathered by bees and produces an acrid, unpleasant honey.

Notes The genus is named for an ancient Greek physician, Euphorbus, physician to the king of Mauritania. The species epithet involves Greek words for cup-bearer, referring to the cup-shaped involucral glands. This plant as illustrated has also been known as *Poinsettia cyathophora* and *Euphorbia heterophylla*.

Candelabra "cactus"   p. 34
Euphorbia lactea   Euphorbiaceae

Toxic properties The milky, sticky latex which runs abundantly from injured stems is caustic to the skin and mucous membranes externally. Internally, it is bitter, irritant, emetic and purgative. It contains several proteolytic enzymes which probably contribute to its irritant properties. Experimentally, the latex shows cocarcinogenic properties.

Symptoms The sap causes a burning rash and or blisters on the skin. In the eye, the sap produces severe irritation with small hemorrhages in the conjunctiva, corneal clouding and corneal ulceration. Temporary blindness may last several days. The maximum effects are usually experienced within the first day and diminish significantly by the third day. Full recovery from the corneal clouding may take 2 or 3 weeks. Internally, the latex produces bleeding and inflammation of the intestinal tract. Fatalities have resulted from the use of the latex in internal folk remedies. The diterpenoids in the resin have shown activity as tumor promoters (cocarcinogens).

Treatment Sap on the skin should be washed off immediately with soap and water. Sap in the eye should be rinsed with clean water before seeking medical assistance. Corticosteroid eye drops help relieve symptoms and speed recovery.

Beneficial uses The plant may have use as a pesticide in underdeveloped countries, as the latex is very toxic to *Lymnaea* snails, which serve as intermediate hosts for several human and livestock parasites.

Notes The genus is named for an ancient Greek physician, Euphorbus, physician to the king of Mauritania. The species epithet is Latin, meaning "milk-white," referring to the sap. This plant is also known by the common names of monkey puzzle, mottled spurge and Malayan spurge.

Crown of thorns

p. 35

Euphorbia milii var. *splendens*   Euphorbiaceae

Toxic properties A series of diterpene esters of ingenol called milliamines are the primary irritants and cocarcinogens contained in the sap. Euphorbol is the primary member of this group. The abundant, sharp thorns produce mechanical injury when people fall into or against the plant.

Symptoms The sap causes severe irritation or temporary blindness in the eyes. Redness, swelling and blisters result from skin contact with the sap.

Treatment Latex on the skin should be removed immediately by vigorous washing with soap and water. The thorns are nontoxic and produce only mechanical injury.

Beneficial uses The plant has been shown to contain lasiodiplodin, which has potent antileukemic activity in laboratory animals. The latex at very low concentration is lethal to the aquatic snails that serve as the vector for schistosomiasis and thus may provide a low-cost, locally produced product to control this dreadful disease in many developing tropical nations. Folk medicine has used the milky sap to remove warts and to stop the flow of blood from cuts. The stem root and latex are used in Chinese medicine to treat hepatitis and abdominal swelling.

Notes The genus is named for an ancient Greek physician, Euphorbus, physician to the king of Mauritania. The species is named for Baron Milius, the one-time governor of Bourbon (now Reunion). This plant has also been known as *Sterigmanthe splendens* and *Euphorbia splendens*. Other common names include Christ thorn and Christ plant after its reputed use as a mock crown on the head of Christ at his crucifixion.

Black manchineel

p. 35

Euphorbia petiolaris Euphorbiaceae

Toxic properties The plant contains several as-yet-unidentified skin irritants.

Symptoms Native people report the sap from this tree produces severe skin irritation and blisters as large as golf balls. The few medical reports indicate the sap may be one of the most potent skin irritants in the family.

Treatment Avoidance of the sap is the best policy. If the sap contacts the skin, it should be removed immediately by washing with soap and water. Because there is no scientific body of knowledge regarding this plant's phytochemistry, treatment should be directed at removing the sap and alleviating the symptoms.

Beneficial uses Folk medicine has used the plant to remove warts.

Notes The genus is named for an ancient Greek physician, Euphorbus, physician to the king of Mauritania. The species epithet alludes to the long petioles, which are as long as the leaf blades. This plant has also been known as *Aklema petiolare*.

Poinsettia

p. 36

Euphorbia pulcherrima Euphorbiaceae

Toxic properties The diterpene esters found in most species of *Euphorbia* seem to be present in only small amounts in most commercially produced varieties of poinsettia. Human and animal irritant response to the plant is inconsistent and seems best explained by the presence of an as-yet-unidentified biologically active sensitizing agent. The recently isolated protease *euphorbain P* may play a part in the irritant response to this plant.

Symptoms The fresh latex causes irritation and blistering of the skin of sensitive individuals, but leaves others untroubled. Latex in the eye causes severe inflammation and sometimes temporary blindness. Internally, small amounts inflame the mouth and throat and cause irritation of the gastrointestinal tract, leading to diarrhea. A report indicates that a 2-year-old child experienced severe vomiting, diarrhea and eventually death after consuming a poinsettia leaf. Other individuals are able to consume the bracts with impunity. Allergic sensitivity to this plant may result in delayed onset of caustic symptoms hours or days after exposure to the latex or an aqueous extract from the leaves.

Treatment The sap should be washed from the skin immediately with soap and water. The latex in the eye should be rinsed copiously with clean water.

Leaves, fruit or sap should be removed from the stomach if vomiting has not occurred.

Beneficial uses This plant is a major article of seasonal horticultural commerce. In contradiction to much of the adverse literature and dire toxic reputation, the Javanese have eaten the leaves as a seasoning. In folk medicine, the latex is used as an emetic and to cauterize aching teeth and poisonous insect bites. The leaves are used in a poultice to treat body aches, fevers and skin diseases. The leaf tea has been used by the Aztecs and modern peoples to promote lactation. The macerated plant is used to poison fish and a red dye has been extracted from the bracts.

Notes The genus is named for an ancient Greek physician, Euphorbus, physician to the king of Mauritania. The species is from the Latin word meaning "most beautiful." This plant has also been known as *Poinsettia pulcherrma.* The vernacular name is for Joel R. Poinsette, the first U.S. minister to Mexico, who introduced the plant to the U.S. in 1833. Other common names include Christmas star, lobster plant, star of Bethlehem and Mexican flame tree. The plant has been subject to considerable manipulation by the horticulture trade, which has produced cultivars with multiple whorls of floral leaves. The plant produces maximum color of the bracts in response to short days; thus, plants growing in the open are the most vivid at the solstice on December 22. The horticulture industry sometimes manipulates day length in greenhouses to induce color development for early season marketing. Analysis of the latex has shown the presence of germanicol, beta-amyrin, pseudotaraxasterol, beta-sitosterol, octaeicosanol and the sterol pulcherrol.

Pencil tree   p. 37
Euphorbia tirucalli   Euphorbiaceae

Toxic properties A fine spray of the latex often is produced when chopping this plant. Cut surfaces also ooze a significant amount of liquid latex. The latex contains ingenane, more than 15 irritant tigliane diterpene esters, triterpene esters and the steroids euphol and tirucallol. There seems to be some chemical differences in races of this plant originating in South Africa and Madagascar.

Symptoms The latex causes a burning rash and or blisters on the skin. In the eye, the latex produces severe irritation with small hemorrhages in the conjunctiva, corneal clouding and corneal ulceration. Fresh liquid latex or dried material introduced by rubbing the eye with contaminated fingers are equally effective at producing symptoms. Temporary blindness may last several days. The maximum effects are usually experienced within the first day and diminish significantly by the third day. Full recovery from the corneal clouding may take 2 or 3 weeks. Internally, the latex produces bleeding and inflammation of the intestinal tract. Fatalities have resulted from use of the latex in internal folk remedies. In laboratory experiments with mice, the latex from this species has been one of the most irritating and most likely to produce tumors of the *Euphorbia* tested. The diterpenoids have shown activity as tumor promoters (cocarcinogens).

Recent research has linked the high incidence of Burkitt's lymphoma in parts of Africa to carriers of the Epstein-Barr virus, which has been activated by exposure to the tumor-promoting 4-deoxyphorbol found in the latex of this plant. The relevance of this epidemiological finding to our industrialized world is that the Epstein-Barr virus has been linked to infectious mononucleosis. If you have had mononucleosis, you probably should not mess with this plant!

Treatment Sap on the skin should be washed off immediately with soap and water. Alcohol aids in the removal of dried material. Sap in the eye should be rinsed with clean water before seeking medical assistance. Corticosteroid eye drops help relieve symptoms and speed recovery. Considering the intense irritant response of the eye to the latex, it would be prudent to wear goggles any time the plant is to be pruned.

Beneficial uses This plant has been used as a hardy ornamental and as a dense barrier hedge effective even against livestock. The latex has been proposed as a source of fuel hydrocarbons, but difficulty of harvest and worker safety considerations have prevented an effective enterprise. The chopped plant has been found to enhance the production of fuel biogas when mixed with cow dung. A derivative of the ingol esters in this plant has been found experimentally to possess cytotoxic activity against leukemia cells in rats. There have been repeated, geographically diverse, unsubstantiated claims that the latex is a cancer cure. An alcohol extract from the twigs has shown activity against *Entamoeba histolytica*. Folk medicine has used the latex externally to treat warts and internally (but dangerously) to treat asthma, colic, diarrhea, headache, hemorrhoids, leprosy, neuralgia, rheumatism, sexual impotence, snakebite, syphilis, sore throat and wounds. The twigs have been used as a fish poison and have been inserted in the vagina to induce abortion. The latex cooked with rice has been used as a poison to eliminate crop-destroying birds.

Notes The genus is named for an ancient Greek physician, Euphorbus, physician to the king of Mauritania. The species epithet is from the Malayalam (an ancient Dravidian language spoken on the Malabar coast of India) name *thirikkalli*, in which *thiri*, meaning "thread," refers to the stems and *kalli* refers to its succulent nature. Other common names include aveloz, milkbush, Indian spurge, naked lady and wishbone cactus. The resin euphorbon is the primary constituent of the latex. Chemically, the diterpenoids in the latex are

4-deoxy-4alpha-phorbol-12,13,20-triacetate and 12-O-2Z-4-E-octadienoyl-4-deoxyphorbol-13-acetate. Also found in extracts from this plant are cycloe-uphornol, euphorbinol, euphorbosterol, euphoron, taraxerone and taraxasterol.

Manchineel   p. 37
Hippomane mancinella   Euphorbiaceae

Toxic properties All parts of the plant are toxic by ingestion or direct skin contact. The plant also induces an allergic dermatitis. The aerosol produced by chopping the wood of a live tree and the smoke from burning leaves or woods are irritating to the skin, eyes and respiratory system. The toxic elements are quite varied and all have not been chemically identified. Some of the toxins are water-soluble while others are soluble only in oil or organic solvents. Some produce an instantaneous irritation while others take some time to make their influence felt. Certain of the compounds seem to be cocarcinogenic and induce formation of benign and malignant tumors. The water-soluble toxic component of the fruit is very similar to and may be the alkaloid physostigmine. There are a series of methyl derivatives of elagic acid in the leaves. A water-soluble compound in an ethanol extract from the leaves named hippomanin forms pale-yellow crystals. The ripe fruits are significantly more delicious and toxic than the green ones.

Symptoms The milky sap is highly irritating to any part of the body. It causes an unpleasant rash and eventually blistering on the skin and pain and blindness (usually temporary) in the eye. It is rumored that rain dripping from the leaves will cause acute dermatitis in humans and kill birds roosting in the tree. Exposure to the smoke of burning wood has produced headache, acute dermatitis, temporary painful blindness and severe respiratory problems. If the fruit is chewed and swallowed, the effects—burning pain, salivation and swelling of the lips, tongue and gums—may be delayed for as long as several hours. Sloughing of the gastric mucosa is usually evident. Abdominal pain, vomiting and bleeding of the digestive tract are usual. An old Danish physician who practiced medicine most of his life in the West Indies recounted that at autopsy, the gastrointestinal tracts of victims of manchineel poisoning look as if they had consumed a can of lye. Shock and death may occur without supportive therapy. Chronic exposure may be lethal due to degeneration of the liver, kidneys, pancreas and adrenals. Although some accounts report individuals who can enjoy the fruits with impunity, I strongly recommend forgoing the experiment.

Treatment For external exposure to the sap, prompt multiple washings with soap and water are the best way to reduce or prevent adverse symptoms.

Some of the proprietary poison oak and poison ivy cleansers may help remove the compounds adsorbed by the skin. Gastric lavage with oil followed by a saline cathartic is a recommended treatment for ingestion. Do not induce vomiting except as a last resort when a patient cannot reach medical facilities in a reasonable time. Epinephrine may be used as an antidote for the fall in blood pressure. Folk medicine uses the root of *Jatropha gossypiifolia* to treat *Hippomane* dermatitis.

Beneficial uses Folk medicine has used extracts of the leaves and bark as a vermifuge and a cathartic, and to treat scabies, syphilis, tetanus, venereal diseases and warts. The attractive, decorative, durable wood takes a fine polish and resembles fine walnut but with more variable color. It has been used for furniture, cabinets, interior trim and construction. Once it is formed and finished, the wood produces no adverse reaction on casual contact. Bees produce abundant nontoxic honey from the flowers.

Notes The generic name is from the Greek word meaning "horse poison." The species epithet is from the Latinization of Spanish words meaning "little apple," for the remarkable similarity of appearance and aroma of the fruits to small apples. Another common name for this plant is beach apple or Spanish manzanillo. Columbus recorded that the Carib Indians used the milky sap to poison arrows. Certain birds and iguanas have been seen to consume the fruit without apparent adverse effects. Land crabs, *Cardisoma guanhumii*, a culinary delicacy in the Caribbean, eat the fruit and leaves and reputedly retain sufficient toxin in their flesh that they have an adverse effect on humans eating them.

Sandbox tree   p. 38
Hura crepitans   Euphorbiaceae

Toxic properties The translucent yellow sap is caustic and poisonous containing the toxic proteins hurin and crepitin which are lymphatic mitogens. The LD_{50} of hurin is reported as 0.2 mg/kg I.V. The physiologically active toxic lectin in the seed has not been identified. The tricyclic diterpene daphnane, huratoxin from the sap is about 10 times more toxic to fish than rotenone. This is said (humorously) to be one of the most dangerous trees in the forest because it stabs, poisons and shoots its victims.

Symptoms The sap causes immediate inflammation and later eruptions on the skin and painful irritation to the eyes, sometimes so severe as to induce temporary blindness. Dust from working the wood or smoke from its burning irritates the eyes and respiratory tract. Consumption of half a pleasant-tasting seed can rapidly produce debilitating intestinal cramps, diarrhea, and vomiting followed by rapid heartbeat and impaired vision. It is reported that the digestive disturbances may be delayed a day or more after consumption of a seed. In large doses, comprised of two or more seeds, delirium, convulsions, and death may ensue. Segments of the woody fruit used in jewelry handicrafts have caused dermatitis.

Treatment Immediate washing of the afflicted part with soap and water helps remove the irritant from the skin. Immediate copious rinsing of the eyes with clean water should be accomplished to remove most of the sap before seeking ameliorating medical treatment. The violent vomiting and diarrhea associated with the consumption of the seeds is usually self limiting but may be life threatening when children consume large amounts of the pleasant-tasting seed. If several seeds are eaten or symptoms are severe, medical care should be sought for replacement of fluids and electrolytes.

Beneficial uses Hurin in very low doses has been found to stimulate mitosis (cell division) by a factor of over 75 times in T (thymus derived) lymphocytes. This effect could have wide therapeutic value in human medicine. The fine-textured wood with a silky luster is soft and light with an interlocking grain. It is used for rough construction, crates, plywood and veneer, and cheap furniture. Before the European invasion the Native Americans made dugout canoes capable of carrying 40 men from the hollowed trunks. The juice from minced bark has been used to disable fish for capture. Smoke from the smoldering wood repels insects but is quite unhealthy to humans. The seeds, seed oil, and leaves have been used in folk medicine as a purgative, sometimes producing considerably more vigorous results than desired. In recent times the seeds have been sold by herbal shops with the intent that they be used as aversive conditioning for problem drinking habits. The roasted powdered seed is included in food or drink and produces vomiting and diarrhea associated with alcohol consumption later in the evening. A leaf poultice has been applied to treat headaches, rheumatism and other pains. The sap has been used as a vermifuge, probably killing the worms but leaving the patient in a less than perky state. It has also been used to treat leprosy and elephantiasis

Notes The generic name is derived from a Native American word meaning "poisonous sap." The species epithet is Latin meaning rustly or crackly. The common name is derived from the historic use of the hollowed immature fruits as a container for fine sand used for blotting ink. The tree is also known as monkey pistol. The explosive dehiscence of the pod flings the seeds at velocities of up to 70 meters per second to a distance of up to 45 m (150 ft) from the parent tree. Large old trees often develop central hollows which are used as a residence by bats and other wildlife. The sap was used as an arrow poison by indigenous peoples. The individual woody sections of the disintegrated fruit are often used in handicrafts. Macaws and other parrots are said to feed eagerly on the seeds.

Physicnut
Jatropha curcas

p. 39
Euphorbiaceae

Toxic properties The seeds contain a dramatically purgative, pleasant-tasting oil called Pinhoen oil, hell oil or oleum infernale and the toxic protein curcin (also called jatrophin), which inhibits protein synthesis in the cells of the intestinal walls, increases prothrombin time and has other adverse influences similar to those of ricin and abrin. The irritating diterpene component of the seed oil has been identified as 12-deoxy-16-hydroxyphorbol. The seeds are strongly cocarcinogenic. The twigs and leaves contain numerous complex chemicals. The bark contains a steroidal sapogenin. The sap, containing the proteolytic enzyme curcain, is irritating externally to the skin and eyes and toxic internally, but rapidly halts bleeding when applied to wounds.

Symptoms The seeds and the oil produced from them rapidly produce burning in the throat and abdominal pain followed by vomiting and diarrhea when consumed in excess. Human deaths have resulted from consumption of the seed. Experimentally, the powdered seed has been found to depress cardiac function, blood pressure and respiration. The oil produces redness and eruptions of the skin when used as a poultice. Roasted seeds are considerably less toxic, but may produce sores in the mouth if eaten in quantity. The toxic internal symptoms produced by curcin may be delayed for some time. Livestock consuming the plant or its seeds may show diarrhea, dilated pupils, staggering, bloat, paralysis, fever, reduced water consumption with dehydration, shivering and coma followed by death in some cases. Postmortem examination shows hemorrhage of the heart, lungs, kidney and spleen. Fatty change and necrosis of the liver is typical. Inclusion of 0.5% of the seed in poultry feed results in kidney and liver pathologies along with internal hemorrhages.

Treatment Dehydration and loss of electrolytes should be corrected.

Beneficial uses Extracts of the leaves have been effective against leukemia in test-tube studies and in experimental mice. The latex has been experimentally demonstrated to be active against *Staphylococcus aureus*. The mucilage from the seed pulp reduces prothrombin time and is used as a source for thromboplastin. Women in Sudan have used small doses of the seeds as an oral contraceptive. The contraceptive effect has been confirmed with laboratory rats. Both humans and rats recovered fertility upon eliminating the seed from the diet. Folk medicine has used a piece of root kept overnight in the vagina to induce abortion.

The oily, pleasantly flavored seeds have been marketed by herbalists under the names of physic nuts, purging nuts, Barbados nuts and several variants in French and Spanish. They are a severe purgative and have caused the death of children. One indistinguishable cultivar of this tree has delicious nontoxic seeds which may be consumed in moderation with no adverse effects. I find them to have a mild, rich flavor similar to that of cashews. The seeds contain 58% protein. The seed has been used in folk medicine to treat burns, convulsions, diarrhea, fevers, gonorrhea, hair loss, incontinence, rheumatism, scabies, sciatica, syphilis, tetanus, upset stomach, whitlows, yaws and yellow fever. The leaf tea is used in folk medicine as a bath to relieve fevers, colds and childhood marasmus. The leaf tea is consumed to relieve venereal diseases, heartburn, fevers, constipation, diarrhea and jaundice and to promote milk secretions in nursing mothers. The young leaves have been cooked and eaten as a vegetable. The sap has been used as a purge, to fill cavities in aching teeth, to arrest bleeding from wounds, to cure burns, hemorrhoids, skin diseases, sores and wounds, and to treat stings of bees and wasps. It is reported that the seeds are tasty and innocuous when roasted, but other reports claim that they are prone to produce sores in the mouth. The oil pressed from the seed has greater purgative activity than castor oil but less than croton oil, and has been abandoned as being too potent. It is used as a lubricant, for soap making, as the base for paints and can be burned as a smoke-free illuminant.

The seed oil may be used as a fuel for diesel engines and is being promoted actively as a renewable, low-cost resource in developing countries. The young leaves are reputed to be safely eaten if cooked. They are said to counteract the peculiar smell of cooked goat meat. The bark has been used to culture scale insects for the production of lac, a component of varnish for musical instruments and other fine woodwork. A leaf extract has been used to set the dye in cotton fabric. The sap, leaf and bark have been used as intoxicants to catch fish. A dark-blue dye has been prepared from the bark. Bees use the abundant flowers as a source for a dark-amber honey with a strong but pleasant flavor.

Notes The generic name is from the Greek words *iatros*, "physician," and *trophe*, "food." The species epithet is Native American. It has also been known as *Curcas curcas*. Other common names include Chinese peanut and wild pistachio. The nuts are carried by some people as pocket charms and are called lucky nut, but they have also been used as the instrument for homicide. The oil from the seeds is composed primarily of stearic, palmitic, myristic, oleanic and curcanoleic acids. Curcanoleic acid is similar to the ricinoleic and crotonic acids found in castor bean and croton oil. The mucilage from the seed pulp is composed primarily of xylose, galactose, rhamnose and galacturonic acid. The flavonoids apigenin, vitexin and isovitexin have been isolated from the leaves.

Wild physicnut
Jatropha gossypifolia

p. 40
Euphorbiaceae

Toxic properties The seeds contain the toxic protein curcin and a purgative oil containing the irritating diterpene 12-deoxy-16-hydroxyphorbol. The toxic, bitter alkaloid jatrophine has been obtained from the roots and bark. The leaves contain histamine and tannin.

Symptoms Consumption of the seeds and the oil produced from them rapidly produces abdominal pain followed by vomiting and diarrhea when consumed in excess. The toxic symptoms produced by curcin may be delayed for some time. The high level of esophageal cancers in the residents of Curacao has been linked to regular consumption of folk medicines containing the cocarcinogenic diterpene esters from this plant. Contact with the plant can produce an acute histamine reaction. The sap sometimes produces dermatitis.

Treatment The symptoms are self-limiting and resolve themselves within 24 hours. Dehydration and loss of electrolytes should be corrected.

Beneficial uses The diterpene jatrophone from the foliage, bark and roots has been found to inhibit the growth of several different cancers. It inhibits leukemia in mice, cell cultures of human nasal cancer and 4 other standard animal tumor systems. Jatrophine extracted from the leaves has been found to be active against malaria parasites in laboratory experiments. The leaves are used in folk medicine to treat malaria. Folk medicine has long used various extracts to treat cancer. In folk medicine, the leaf tea is taken as a laxative and is used to treat fevers, hepatitis, diarrhea, constipation, ulcers, colic, diabetes and venereal disease. Externally, it is applied to sores and rashes. The seed oil is used to treat leprosy. The sap is applied to burns, ulcers and sores in the mouth. It effectively stops the bleeding and discomfort of scratches. The roots are used to treat kidney, liver and bladder problems and are considered a remedy for snakebite, leprosy and the dermatitis caused by manchineel. The root tea is given to animals to promote healing of bone fractures. Sap from the leaves is used as a flux in mending iron pots and as an intoxicant for fish capture. Oil from the seeds is used in lamps. The sap is toxic to various parasite- and disease-carrying snails. A methanol extract of the fruit is toxic to the snail *Bulinus globulus* at the concentration of 12 ppm. Birds actively seek and consume the seeds.

Notes The generic name is from the Greek words *iatros*, "physician," and *trophe*, "food." The species epithet is Latin, referring to leaves that look like those of cotton, *Gossypium*. Other common names are tattoo bush, African coffee and bellyache bush.

Coral tree p. 40
Jatropha multifida Euphorbiaceae

Toxic properties The sweet, pleasant-tasting seeds contain the toxic protein curcin (also called jatrophin), which inhibits protein synthesis in the cells of the intestinal walls and has other adverse influences similar to those of ricin and abrin. The seeds yield 30% to 40% of a potent purgative oil whose active ingredient is 16-hydroxyphorbol. Also present are various terpene cocarcinogens. The leaves and sap seem to have similar compounds in lesser concentration. The sap is an irritant and causes inflammation of the skin and eye. It contains two immunologically active acylphloroglucinols. The leaves contain a saponin.

Symptoms Consumption of a single seed or the ripe fruit will cause vomiting and purging lasting from 8 hours to as long as a week, often accompanied by dilation of the pupils. The toxic symptoms produced by curcin may be delayed for some time. Livestock consuming the plant or its seeds may show dilated pupils, staggering, bloat, paralysis, fever, shivering and coma followed by death in some cases. Severe poisonings and deaths have been recorded from the effects of the seeds. The adverse effects of this plant are generally more severe than those of *Jatropha curcas*.

Treatment Dehydration and loss of electrolytes should be corrected. Lime juice and stimulants are used as an antidote in folk medicine.

Beneficial uses The tuberous roots are reputed to be roasted and used as a starchy vegetable. Folk medicine uses the seeds, their oil and the tea prepared from them as an emetic and a purgative. The oil kills lice and reputedly the mites which cause mange. The sap is applied to cauterize and stop hemorrhage in wounds, to reduce the swelling in bruises and abscesses and to treat scabies and other skin diseases. Folk medicine claims that the seed oil produces abortion.

Notes The generic name is from the Greek words *iatros*, "physician," and *trophe*, "food." The species epithet is Latin, referring to the many splits in the leaves. The horticulture trade has marketed this plant as star of India, nutmeg plant and Chinese umbrella tree. Other common names in Spanish, French, Dutch and English include variations on its use in folk medicine as an emetic, physic or purging agent.

Cassava
Manihot esculenta

p. 41
Euphorbiaceae

Toxic properties   The tubers contain the cyanogenic glycosides linamarin (93%) and lotaustralin (7%) which undergo acid hydrolysis in the digestive tract to release cyanide (hydrocyanic acid). Hydrocyanic acid interferes with the chemistry of respiration and prevents the utilization of oxygen by the cells, leading to death by asphyxiation. The minimal lethal dose of cyanide for humans is about 50 mg. With a range of cyanide concentrations from 30 to over 450 mg/kg in the roots, a lethal dose could be acquired with the consumption of 125 g (4 oz) of the untreated tuber. Indigenous peoples have often classified the plants into bitter (toxic) and sweet (nontoxic) forms. The forms with less than 0.01% hydrocyanic acid are usually considered to be sweet. All forms must be peeled before consumption as the toxin is always more abundant in the skin. Botanists have found no consistent morphological or ecological characteristics which correlate with cyanide content. The cyanide content of the tubers does vary greatly depending on cultivar, soil nutrients, soil moisture, age of the plant and growing conditions. Leaves cooked in a traditional manner release the mutagenic compound quercetin.

Symptoms   Symptoms may be delayed several hours after consumption until hydrolysis and absorption have placed the toxin in the bloodstream. Vomiting, diarrhea, headache and abdominal pain may be followed by lethargy, dizziness, labored breathing, chills, visual disturbances, sweating and convulsions. Cyanotic symptoms are not inevitable. In severe cases convulsions with intermittent muscle flacidity may lead to coma and death. Chronic cyanide poisoning in some African populations due to the regular consumption of the inadequately treated tubers has lead to cretinism, goiter, and ataxic tropical neuropathy. Lethal poisoning has been reported in horses, cows, sheep, and most frequently, in pigs which greatly relish the tuber.

Treatment   The stomach contents should be immediately removed. Activated charcoal will adsorb the cyanide but release it slowly in its passage through the intestinal tract. Several hundred ml of 25% sodium thiosulphate solution will neutralize most of the cyanide in the digestive tract by reacting with it and producing thiocyanate. In emergencies the oral administration of this chemical obtained from photographic darkrooms has saved lives. A physician should be prepared to treat for shock, provide oxygen and respiratory assistance, and provide intravenous cyanide antidote. Chronic cyanide poisoning is usually associated with protein malnutrition and extremely low levels of sulfur amino acids in the blood. Treatment of the nutritional deficiencies is the preferred method of treating affected population groups.

Beneficial uses The cooked tuber is the major source of calories in many tropical societies and it has been estimated that half of the world population depends on cassava. The "sweet" forms are usually boiled or baked. The "bitter" forms may be grated and washed in several changes of water to eliminate the cyanide before the residual starch mass is squeezed free of water and baked into bread. More of the tuber is consumed in Africa than in its native Latin America. Cassava provides 10 times more calories per unit area of crop than maize. The root contains about 135 calories per 100 gram serving and has significant amounts of calcium, phosphorus, niacin, and ascorbic acid. The cooked, dried, and ground tuber is marketed as cassava meal and a specially prepared starch extract is the tapioca of commerce. Tapioca is used in the baking industry and in the brewing of a superior beer. The extracted starch with a high viscosity and considerable tensile strength when dry is used in the laundry, paper and textile industries. A glue suitable for postage stamps has been prepared from the starch. Fermentation followed by distillation yields alcohol suitable for use in beverages or as a motor fuel. The leaves are rich in protein and vitamins and have been recommended as dietary supplements for nutritionally deprived rural people. The B vitamin content of young leaves is such that they have been recommended as a treatment for beri-beri after detoxification. The leaves are widely consumed in Africa after detoxification by chopping and boiling. The poisonous sap reduced to an innocuous antiseptic syrup by boiling is called cassareep and is used as a meat preservative. The chronic cyanide poisoning associated with habitual consumption of the tuber confers an advantage in reducing the symptoms of sickle cell anemia. Various parts of the plant have been used in folk medicine to treat angina, boils, cancer, diarrhea, eczema, hepatitis, rheumatism, tumors, scabies, snakebite, sores, and whitlows. The peelings are an effective fish poison.

Notes The generic name is derived from the original Brazilian native name, *manioc*. The species epithet is Latin meaning "good to eat." This plant has also been known as *Manihot utilitissima*. The common name is from an original native name, cazabe, which has been perverted in several steps to the present cassava in English. Other common names include manioc (derived from the Guaraní name, but used today mostly in French speaking areas), yuca (from the Arawak name), tapioca (in Asia), and sweet potato. In experiments with laboratory animals fed a diet containing levels of cyanide similar to that ingested by some human groups using cassava as a major food source, significant behavioral differences were noted: An increasing ambivalence and slower response time to stimuli were coupled with an energy conservation behavior in which vigorous activity was avoided.

Christmas candle
Pedilanthus tithymaloides

p. 42
Euphorbiaceae

Toxic properties The very caustic, milky juice of the roots, stems and leaves contains euphorbol and other diterpene esters which are irritants and cocarcinogens. The presence of a lectin and of proteolytic enzymes is experimentally indicated.

Symptoms If ingested, a few drops of the juice produce irritation of the mouth and throat, vomiting and diarrhea. Externally, the juice produces irritation, inflammation and blistering of the skin. The lesions on the skin of livestock are prone to secondary infections. The sap produces an intensely painful irritation of the eye, often followed by keratoconjunctivitis and temporarily reduced visual acuity. The seeds cause violent, persistent vomiting and drastic diarrhea.

Treatment Avoidance of contact with the sap is the best policy. It has been recommended that protective goggles be worn when cutting the plant. Immediate washing with soap and water is the best treatment for accidental skin contact. Copious rinsing with fresh water, followed by professional medical attention, is recommended for eye contact with the sap. Topical steroids often help reduce pain and inflammation while recovery is proceeding. Fluid replacement may be needed in acute cases of internal consumption. Due to the irritation and unpleasant taste of the sap, livestock seldom consume enough to require veterinary care.

Beneficial uses The flowers are particularly attractive to hummingbirds. A proteolytic enzyme called pedilanthain extracted from the latex is experimentally effective against intestinal worms and reduces inflammatory reactions when taken orally. Folk medicine has used this plant as an unsubstantiated and probably dangerous cure for numerous maladies. The leaf tea has been used to treat laryngitis, ulcers of the mouth, venereal disease, asthma and coughs. The root tea has been used to induce abortions and as a purgative substitute both in name and function for ipecacuanha. The latex has been used to treat cancer and umbilical hernia and to drip into decayed aching teeth and aching ears. Its caustic nature might give it some efficacy in the traditional use in treating warts, calluses and ringworm. The arrangement of the leaves in rows similar to centipede legs may account for the persistent and widely occurring folk use as a treatment for centipede and scorpion stings.

Notes The generic name is from the Greek words *pedilon,* meaning "slipper," and *anthos,* meaning "flower." The species is named for its resemblance to plants in the genus *Tithymalus.* The plant has also been known as *Euphorbia tithymaloides.* Other common names are fiddle flower, devil's backbone, slipper flower, Japanese poinsettia, redbird flower, redbird cactus and ipecacuanha.

Gale-of-wind
Phyllanthus niruri

p. 43
Euphorbiaceae

Toxic properties The plant has many physiologically active alkaloids in the fruit, leaves and roots. Included are dhurrin, a cyanogenic glycoside, a bitter crystalline substance phyllanthin in the bark, and a catechol carboxylic acid.

Symptoms The foliage of the plant or a tea made from it induces diarrhea. Some individuals have shown symptoms of cyanide poisoning after drinking the tea.

Treatment The intestinal disturbance is self-limiting, but a patient should be monitored initially for the effects of cyanide poisoning.

Beneficial uses In laboratory tests, extracts of this plant showed activity against hepatitis B virus. The compounds phyllanthine, hypophylanthine and triacontanal, extracted from the whole plant, have shown protecteive effects against liver damage by several toxins. An alcoholic extract of the leaves has been shown to reduce blood sugar in experimental rabbits with an effect comparable to that of tolbutamide. The extract is antibacterial, anti-fungal and antiviral and is effective against *Entamoeba histolytica*. Up to 1 gm/kg of the extract is tolerated orally by the mouse. The macerated plant is used in the Pacific as a fish poison. The plant tea is used in folk medicine as a diuretic, to induce abortion and to treat bladder problems, kidney stones, asthma, colds, colic, constipation, diabetes, dysentery, edema, fever, gonorrhea, influenza, jaundice, kidney and liver maladies, malaria and syphilis. The root is used to treat jaundice or is made into a paste taken for 2 days to induce abortion. The bitter fruit is used to treat bruises, ringworm, scabies, sores and ulcers. An extract of the leaves and stem has been used as an ink and a black dye. The related *Phyllanthus acidus* produces the edible gooseberry, which is highly esteemed for making jams and jellies.

Notes The genus name is from the Greek words for "leaf" and "flower," referring to the fact that in some species the flowers are produced on the edges of leaf-like branches. *Niruri* is a Malayalam name for herbaceous *Phyllanthus*. The plant has also been known as *Phyllanthus lathyroides*. There is considerable confusion in the literature between this plant and *Phyllanthus amarus*. Other common names include Creole senna and poor man's quinine. The flavonoids astralgin, ellagic acid, quercitoside, quercitrin, isoquercitrin, kaempferol and kaempferol 3-glucoside have been isolated from the leaves. The plant also contains 4 leucodelphinidine alkaloids and the lignins phyllanthine, hypophyllanthine and quercitin.

Castor bean

p. 43

Ricinus communis

Euphorbiaceae

Toxic properties The entire plant is poisonous, but ricin, the primary toxic protein, is most concentrated in the pleasant, nut-like tasting seeds. Symptoms may occur within minutes, hours or even days after consuming seeds. Consumption of a single seed may be fatal, but 20 is a more typical lethal adult dose. Ricin, with a molecular weight of 66,000, binds to body cells and disrupts protein synthesis by the prevention of elongation of peptide chains. The toxic effects of injected ricin are so severe that a homicide resulted when a 1.52-mm (0.06-in) platinum sphere with a 0.28-mm hole containing ricin was injected into the thigh of an individual involved in foreign intrigue. Experimentally, a lethal dose injected into a mouse is 0.000001 g. Castor oil, the traditional purgative extracted from the seed, is a triglyceride of ricinoleic acid called ricinolein, which produces its action by stimulating peristalsis in the small intestine. The usual adult dose of 15 ml moves through the intestinal tract so quickly that little is absorbed. The oil typically does not contain any ricin. The seedcake remaining after oil extraction may poison livestock if inadequately heat-treated to destroy the water-soluble ricin.

The oil, seeds and pollen have all produced severe allergic reactions quite separate from their other toxic properties, due to a component called castor bean allergen (CBA). Repeated exposure to the allergen increases the individual's sensitivity to the compound. Many individuals lose the hypersensitivity over a short period of time, but several cases are on record of individuals showing strong allergic reactions to the inclusion of castor oil in lipstick, cosmetic cream formulations and make-up remover. The alkaloid ricinin is also present in the leaves and seeds.

Symptoms Consumption of the seeds may produce symptoms within minutes or with delays as long as hours or rarely days. Symptoms may include burning of the mouth, throat and stomach, abdominal pain, vomiting, bloody diarrhea, extreme thirst, fast, weak pulse, impaired vision, headache, convulsions and blood pressure falling to lethal levels. Clinically, the effects of ricin usually are manifested in the deterioration of the liver, pancreas, spleen and kidney, but may occur as retinal hemorrhaging or disturbance of other organ functions. Ingestion of a single seed of this plant should be considered dangerous, but the critical dose varies by a factor of as much as 100 between individuals, both human and animal. Urinary excretion of ricin is slow, with a half life of 8 days in the body. Patients who survive may be very ill for as many as 10 days before showing significant signs of recovery. Horses and mules are particularly sensitive to ricin and as few as 6 seeds included in their feed may be fatal. Horses may show a profuse, watery diarrhea, sweating, muscle spasms and a strong heartbeat which visibly shakes the body.

Poultry are often poisoned when the seedpods of plants near the run dehisce and scatter the seeds in the run. Large mortalities in flocks of wild ducks have been attributed to consumption of castor bean seeds.

Cattle, sheep and pigs—after they have been poisoned by grazing on the foliage, picking the seeds off the ground or being fed inadequately treated press cake—show a severe gastrointestinal upset, with vomiting, abdominal pain, a watery, sometimes bloody diarrhea, shivering, fever and dilation of the pupils. Cattle and sheep eating the leaves or seeds may show neuromuscular symptoms, including muscle tremors, a swaying gait and chewing movements, perhaps due to the alkaloid ricinin. Pigs may show an uncoordinated gait and weakness. Poultry show drooping wings, ruffled feathers and greyish combs and wattles. The blooming plant may produce asthma, sneezing and eye irritation in sensitive individuals. Allergic skin reactions, conjunctivitis, headache and bronchial asthma have resulted from contact with the seeds used in jewelry or from the use of products such as cosmetics containing castor oil. A high eosinophil count is often associated with chronic allergic response. The standard medicinal dose of 4 to 16 ml of the purified oil reliably produces purging.

Treatment The stomach contents should be removed by syrup of ipecac or other methods. This should be followed by cathartics and activated charcoal, which binds ricin quite effectively. Liver, kidney, pancreatic, heart and red blood cell functions should be carefully monitored. Sedatives and treatment for shock may be needed. Maintenance of a generous alkaline urine flow with 5 to 15 g per day of sodium bicarbonate promotes renal excretion of the toxin while preventing hemoglobin precipitation in the kidney tubules. Fluid and electrolyte replacement are often required, along with alternative nutritive sources in prolonged cases. Repeated small doses of ricin induce a mammalian subject to develop antibodies which provide immunity to the toxin. Serum from hyperimmunized animals neutralizes ricin and provides passive immunity in poisoned animals if supplied promptly after exposure to the toxin. There is no commercially available source of antibodies for treatment of ricin poisoning.

Beneficial uses The extremely toxic nature of ricin has been utilized experimentally to medical advantage. It is linked to monoclonal antibodies which recognize T lymphocytes, and it is hoped that this combination will suppress T cells and prevent the graft-versus-host reaction which causes failure of some organ transplants. The tendency of ricin to bind to various nerve tissues makes it a useful tool in many aspects of neuromedical research. The seeds yield up to 50% of the castor oil which was once used as a laxative but is now used primarily as a high-viscosity industrial lubricant, with world production exceeding 800,000 tons per year. It is also used in formulating hydraulic fluid, inks, transparent soaps, lipstick, hair dressings, synthetic fibers and contraceptive foams, jellies and creams. The oil is used widely in the textile industries as a carrier and fixative for dyes. The plant fiber combined with 30% bamboo pulp is suitable for making paper and wallboard. The press cake remaining after oil extraction contains 20% protein and can be heat-treated to destroy the toxin, then fed to poultry and livestock or used as a high-nitrogen fertilizer. Heat treatment does not alter the allergic properties of the seed, seedcake or oil.

The seed oil was used by the ancient Egyptians as a salve, and the seeds have been preserved with 4,000-year-old mummies. It has been used widely in folk medicine as a purgative, as a tonic and to treat skin problems. Castor oil has been used dangerously to treat arthritis, asthma, burns, cancer, cholera, convulsions, corns, epilepsy, gout, guineaworm, moles, rheumatism, tuberculosis, urethritis, venereal diseases and warts. Fresh leaves are used as poultices to relieve rheumatism, headache, stomachache and fever and as a poultice to stimulate milk flow. The leaf tea is used externally to clean sores and shrink hemorrhoids and is consumed to treat venereal disease. A thorough discussion of the folk uses of this plant would occupy a small book. Bees gather a sweet exudate from extra-floral nectaries from the fruiting stalk of the castor bean plant.

Notes The generic name is Latin, meaning "tick," for the tick-like appearance of the seeds. The species name is from the Latin word meaning "common." In Spanish, it is often called *higuerito* or variations on that name. Castor oil is composed primarily of ricinoleic acid with lesser amounts of dihydrosteric, linoleic, oleic and stearic acids.

Mammey apple
p. 44

Mammea americana
Guttiferae

Toxic properties The active toxic component of this plant is mammein, one of many coumarins found in the seed. The pale-yellow latex from the bark, the powdered seed, the leaves and the flowers are all insecticidal. The seed extract placed in water is lethal to fish, producing symptoms similar to those of phenols. The xanthones in the wood produce a dermatitis in woodworkers.

Symptoms The seeds are toxic to livestock and fowl if consumed. They are particularly attractive to swine. The powdered seeds in water suspension are lethal in contact or when eaten by many species of insects.

Treatment The best way to prevent accidental consumption of the seeds by livestock is to gather the ripe fruit for human use. The ground or whole seeds should be stored in sealed containers.

Beneficial uses The attractive, strong, hard, heavy, reddish-brown wood with an interlocking grain is used for pilings, construction and carpentry. The bark is used in small-scale tanning. The fruits are in considerable demand. They are eaten raw and in fruit salads or are made into marmalade or preserves. The cooked fruit is very resistant to spoilage. The flowers are distilled in the West Indies to produce an aromatic liqueur called *eau de creole* and an essential oil used in perfume. The latex and the powdered seeds have been used as insecticides, to extract chiggers from the skin and to kill ectoparasites on domestic animals. Mammein has shown significant antitumor activity against sarcomas in the laboratory. The seed oil contains compounds which have shown experimental antifungal activity against *Penicillium notatum*, *Candida albicans* and *Aspergillus niger*. The ground seeds have been used to stupefy and capture fish and have been mixed with coconut oil to kill lice. Folk medicine has used the leaf tea to relieve high blood pressure and cure malaria. The bark tea has been used to treat coughs and skin diseases of domestic animals.

Notes The generic name is derived from the original Native American name for the tree. The species epithet refers to the plant's American origin. The volatile components of the fruit have been studied and found to be composed principally of beta-ionone and about 10 other compounds with flowery or fruity odors. The juice from the seeds leaves a permanent stain on fabrics. While some authorities suggest eliminating this tree due to its potential danger to livestock, there was a court case in the U.S. Virgin Islands in which the plaintiff charged a neighbor with depriving him of a significant income as well as enjoyment of life when the defendant cut down a mammea tree near their mutual property boundary. The court ruled in favor of the plaintiff.

Blood root
p. 45

Lachnanthes tinctoria   Haemodoraceae

Toxic properties The plant contains several closely related unique compounds, but the chemicals responsible for the toxic symptoms and the narcotic effects described below have not been identified. A nonglycosidic 9-phenylphenalenone and other aglycones from the root have been found to photosensitize bacteria to the extent that they are killed by light.

Symptoms Native Americans are said to have used the plant to produce mental exhilaration, with a brilliant and fluent speech coupled with a bold and heroic attitude. The pupils are often dilated and the cheeks reddened. The aftereffects were dizziness, headache and grumpiness. A much-repeated story is that black pigs may eat the root without harm, but white pigs have their hooves fall off and after death are found to have pink bones. The latter is certainly consistent with one of the toxins being a photosensitizing compound to which black pigs are immune due to their pigmentation. The toxic syndrome has been reported as similar to that of congenital porphryia.

Treatment In humans, the remaining plant material should be removed from the stomach. It would seem prudent in our present state of ignorance to keep the exposed individual in low-light conditions for several days, then closely monitor blood chemistry and circulating liver enzymes as exposure to light resumes. It is suggested that pigs provided with a good diet in a shaded pen may be able to recover from the toxin.

Beneficial uses Sandhill cranes eagerly feed on the rhizomes with no apparent adverse symptoms. Folk medicine has used the plant to treat coughs, fever, laryngitis, pneumonia, rheumatism, typhoid and typhus. The root has been used as a red dye. This plant is a wetland indicator species.

Notes The generic name is from the Greek words *lachne,* meaning "down" or "wool," and *anthos,* meaning "flower," in reference to the woolly flowers. The species epithet is the Latin word meaning "weed for dyeing." Other common names are red root, paint root and dye root. This plant has also been called *Gyrotheca tinctoria* and *Lachnanthes caroliniana.* The root contains lachnanthoside, lachnanthofluoren, lachnanthocarpone, di-, tri-, and tetramethoxyphenylnapthalides, and 4 hydroxy-3methoxy-5phenyl-1,8-napthaline-anhydride.

Lion's tail
Leonotis nepetifolia

p. 46
Labiatae

Toxic properties A number of labdane diterpenes and phenolic compounds are found throughout the plant. Nepetaefolin is the most abundant and presumed to be the primary source of toxicity. There are two phenolic substances and traces of an alkaloid in the leaves.

Symptoms The leaves have been chewed or smoked for their intoxicating or euphoriant effects. The root, eaten or taken as a tea, is said to produce giddiness or drowsiness. The leaves have been shown in experiments to be toxic to chicks. The plant has been shown to be toxic to rabbits and sheep when provided as the primary dietary ingredient, but there is some doubt about the species of plant tested. The leaf hairs produce a burning rash on some people and the pollen is a respiratory irritant.

Treatment As with many toxicants of this nature, there is no known antidote, but removal of contact with the plant results in recovery from its effects.

Beneficial uses The leaves are fed as a diet supplement to rabbits and the dried spherical fruiting structures are used in ornamental handicrafts. An ethanol extract has shown experimental anticancer activity. Folk medicine has used the root tea to treat the virus causing dengue fever. The leaf tea, in various formulations, has been used to treat asthma, biliousness, coughs, diarrhea, elephantiasis, epilepsy, fevers, impetigo, intestinal worms, leprosy, menstrual problems, paralysis, scabies, snakebite, syphilitic ulcers, tuberculosis, typhoid, yaws, as a diuretic and as a vaginal douche to stop uterine hemorrhaging. The crushed roots are used to stimulate the flow of milk from the breast when it swells and fails to allow milk to flow from the nipples. The root has also been used to treat convulsions. The plant has some reputation for treatment of malaria, but the seeds are the only part showing feeble antimalarial activity in laboratory experiments. The seed extract is very active against *Alternaria* and *Aspergillus* fungi. The powdered seeds are reported to be effective in destroying lice.

Notes The generic name is from the Latin words *leo* and *otis*, meaning "lion's ear," referring to the sharp, pointed floral bracts which stand erect like the ears of an alert feline. The species epithet suggests leaves resembling those of the plant *Nepeta*. Other English common names are bald head, hollowstalk, lion's ear, pompon and rabbitfood. In Spanish, it is known as *molinillo, boton de cadete* and *quina de pasto*. Additional toxic substances present in this plant are the diterpenes leonotin and nepetaefolinol, along with the phenols nepetaefuranol and nepetaefuran.

Jequirity bean   p. 46
Abrus precatorius   Leguminosae

Toxic properties The primary toxic component is the toxalbumin abrin. With a lethal injected dose being 0.005 g (0.00018 oz), this is one of the most toxic materials known. It is about 100 times more toxic when injected than when ingested. Also present are a glycoside, abralin and an amino acid, abrine. The coat of the mature dry seed is so tough and water-resistant that intact seeds accidentally consumed by children usually pass uneventfully through the intestinal tract. As the outer layer of the seed coat is impermeable to water, the bright, water-soluble pigments of the seed remain intact unless the seed coat is breached. Thus, if intact, brightly colored seeds are found in the feces, it is unlikely that significant toxin leached into the digestive tract. The ingestion of a single well-chewed seed may be fatal. The toxin inhibits protein synthesis in cells by preventing elongation of peptide chains. The toxin first interferes with the cells growing in the intestinal wall, then eventually lodges in the liver and disrupts many of its functions. Antibodies to the toxin may be produced by repeated sublethal injections.

Symptoms The first indications of abrin poisoning are delayed for several hours or (rarely) for several days. Loss of appetite, nausea, vomiting, diarrhea and abdominal pain are usually present, along with rapid pulse, cold perspiration and trembling of the hands. Retinal hemorrhages have been recorded. The lining of the intestinal tract ceases to function and has many bleeding patches. Bloody stools may continue for a week or more after the acute phase of intoxication is passed. In livestock, a reduced appetite, reduced water intake, seeming abdominal pain and bloody diarrhea may lead to respiratory distress and deep depression prior to death.

Treatment Immediate gastric lavage or syrup of ipecac is potentially important, but due to delayed symptoms it is frequently too late to be of much help. Cathartics are not usually useful due to the typical vigorous diarrhea and the necrotizing effects in the gastrointestinal tract. Dehydration and electrolyte imbalance should be corrected. There is no specific human treatment for the toxin, but alkalinization of the urine has been recommended to prevent precipitation of hemoglobin and its products in the kidney tubules. Administration of calcium gluconate and arecoline have been suggested as helpful. Merck has developed an antiserum marketed as Antiabrin or Jequiritol for use in livestock.

Beneficial uses Abrin injected in sublethal doses has been found to completely suppress Ehrlich ascites tumors in experimental animals. Abrin stored for several months at 4°C has been found to stimulate mitosis in T (thymus-derived) lymphocytes in very low doses. This effect could have wide therapeutic value in human medicine. The leaves contain up to 10% of the triterpene glycoside glycyrrhizin (the active flavor principle in licorice) and the roots up to 1.5%. Both have been employed as a licorice-flavoring agent. A sweetening agent called abrusside, up to 100 times as sweet as sucrose, has been extracted from the leaves. Because abrin is destroyed by cooking, the nutritious seeds, with 21% protein and 3% fat, have been eaten with impunity by certain cultural groups and in famines.

The plant was established by man in Africa for its decorative seeds at the time of the pyramids and was grown in Europe by the sixteenth century. The seeds have been widely used as beads in necklaces and rosaries and as the rattles in maracas and in similar children's toys. Folk medicine has used the leaf tea or infusion to treat coughs, influenza, intestinal worms, malaria, sore throats, tuberculosis and fevers. The leaves are chewed to treat hoarseness. Root preparations have been used as an aphrodisiac, to treat sore throat and rheumatism and as a diuretic. Other formulations of the plant have been used to treat cancers of the face, mucosa, vagina and vulva. A seed infusion has been used successfully to treat trachoma, but is not recommended, as the resultant irritation sometimes results in permanent loss of sight. The seed infusion is used by many unrelated primitive peoples to treat various eye irritations. The powdered seed is reported to be used as a female oral contraceptive by Ayurvedic physicians and several Central African tribes. A single 0.2 g dose is reputed to be effective for 13 menstrual cycles. Science has shown experimentally that a methanol extract of the seeds produces an irreversible impairment of human sperm motility. A suppository of the crushed seeds is said to induce abortion. Other groups use an alcohol extract of the seeds subsequently evaporated to dryness and the residue inserted in the vagina to induce abortion. Experimental evidence shows that the extracted lipids from the seeds produce uterine contractions. Ayurvedic medicine considers the roots and leaves useful in treating asthma, tooth decay and fever and as a general tonic.

Notes The generic name is from the Greek word *abros,* meaning "delicate" or "soft." The species epithet is from the Latin word *precatio,* meaning "prayer," probably due to its seeds being used as prayer beads. The common name jequirity is from the Tupi-Guarani language of Brazil. This vine is known by many other names, including black-eyed Susan, crabs' eyes, jumbi bead, rosary pea and weather plant, in reference to the leaves closing on cloudy days. History records that these small, hard, uniform seeds were used as a unit of weight for measuring precious metals in Asia in the Middle Ages, with 16 beans comprising a ducat and 64 a shekel. Other compounds present in the plant are abraline, abrine, abrusic acid, campestrol, cycloartenol, gallic acid, hypaphorine, precatorine, squalene and trigonelline.

Angelin
p. 47

Andira inermis
Leguminosae

Toxic properties The bark and seeds are poisonous and lethal in large doses due to the alkaloids berberine and andirine. The latter is also called angelin, which is N-methyltyrosine. The wood contains chrysarobin, actually a series of anthraquinone derivatives.

Symptoms
Ingestion of large amounts of the bark or seed extract produces violent vomiting and purging, and may be fatal in humans. A delirium and narcosis often accompany the gastrointestinal symptoms. Chrysarobin from the wood, used in folk-medicinal mixtures, may cause severe gastroenteritis or kidney damage. When applied to large areas of skin, a sufficient quantity may be absorbed into the body to produce kidney irritation. Woodworkers report skin irritation and narcotic effects from the wood dust. The smoke of the burning wood is injurious to the eyes. The bark crushed in water has been used as a fish poison.

Treatment The toxic effects are usually self-limiting, but rehydration and other supportive therapy may be needed in severe cases. Folk medicine has used lime juice and/or castor oil to treat overdose of the seeds.

Beneficial uses Chrysarobin, extracted from the wood as a yellow-brown powder, has a long history of being prescribed as a fungicide to treat skin diseases such as ringworm. In folk medicine, the bark and leaves, powdered or as a tea, have been used to expel worms, as an emetic or purgative, and to treat eczema, fevers, malaria, psoriasis, mental diseases and yaws. The bark and leaves have also been taken for their narcotic effects. The leaf tea is used to treat *Comocladia* poisoning. The powdered bark has been used as an insecticide and to stupefy fish for capture. The attractive, durable, hard, heavy heartwood is highlighted with light and dark bands from yellow to brown to red. It is used decoratively in fine furniture, flooring and cabinetry, and in the handles of billiard cues, canes and umbrellas. More utilitarian uses include construction carpentry, pilings, bridge timbers and boat building. The pods are much favored by fruit bats such as *Artibeus,* which carry them about in the forest before roosting to gnaw the fleshy pulp. Parrots also feed on the fruit in the tree. Agoutis, rats and other terrestrial animals also disperse the seeds as they carry the fallen fruit and eat it. Bees make heavy use of the nectar to produce a high-quality honey when the tree is in bloom. Butterflies and hummingbirds make vigorous use of the nectar while the tree is in flower.

Notes *Andira* is a Latinized form of a native Brazilian name which refers to the use of this tree by bats. The species epithet is a Latin word meaning "unarmed," referring to the lack of thorns on this species. Other common names are almendro, partridgewood and cabbage bark. Stigmasterol, formononetin, pseudobaptigenin, genistein, daidzein, taxifolin and the two pterocarpans 3-hydroxy-8,9-methylenedioxypterocarpan and 3-hydroxy-9-methoxypterocarpan have been isolated from the tree.

Coffee senna   p. 48
Cassia occidentalis   Leguminosae

Toxic properties The entire plant is toxic. The purgative oxymethylanthraquinone is in the fruit, leaf and root. The seeds contain a toxalbumin and chrysarobin, both of which cause kidney and liver damage unless the toxic properties are destroyed by heat or formalin. The cathartic action of the seed is produced in part by the presence of emodin. An unidentified toxic alkaloid has also been reported. A toxin which causes muscle degeneration is most abundant in the seed, but is also present in the leaves and stems. The toxin has not been chemically identified, but experiments with rabbits showed the seed toxin disrupts the mitochondria in cells, resulting in muscle degeneration, particularly in the heart. Meat from poisoned cattle may contain sufficient residue of toxin that dogs feeding on it are temporarily incapacitated.

Symptoms A single exposure by ingestion of the raw seeds or a tea made from them produces a self-limiting diarrhea in man. Fever is absent in *Cassia* intoxication. Cattle are more inclined to eat the plant in a pasture a few days after a frost. The seeds and foliage retain their toxicity when dry and have caused toxicity when included in hay. Cattle, horses, sheep and goats consuming the foliage of the plant show diarrhea within 2 to 4 days, sometimes followed by constipation several days later. Although it is not uncommon for animals to continue to eat until hours before death, severe cases of chronic ingestion often lead to loss of appetite. Other typical symptoms include muscle degeneration, tremors, stiffness, a stumbling, swaying gait, a fast, weak pulse and labored breathing, with kidney and liver damage producing a dark-red or brown urine. Afflicted animals commonly remain alert and continue to eat. Clinical signs or inability to rise often only briefly precede death at 5 to 7 days after ingestion.

Postmortem examination reveals hemorrhage and congestion in the heart, intestines, liver, lungs, spleen and kidney and a characteristic pallor or striated pattern on the muscles. The ultimate cause of death is usually degeneration of the heart muscle. Roasted seeds show about half the toxicity of fresh seeds when ground and fed experimentally to goats. Rabbits and sheep have also been recorded as consuming lethal amounts of the foliage of this plant. Pigs have been reported with similar symptoms after *Cassia* seeds were accidentally included in their feed. In chickens, when the seed makes up 0.5% of the diet, mortality in the flock increases; when the seed makes up 4% of the diet, mortality reaches 65%. Chickens show muscular weakness, hypothermia, loss of body weight, ruffled feathers and lack of response to stimulation before death.

Treatment For livestock, the best treatment is prevention of consumption by removing the plant from pastures. The prognosis for recovery is good for animals which avoid or recover from recumbency. Although symptoms of intoxication resemble those of vitamin E-selenium deficiency syndrome, the administration of vitamin E or selenium is contraindicated, as they seem to synergize and enhance the toxicity.

Beneficial uses The roasted mature seeds have been ground and used to extend or substitute for coffee, although the seed contains no caffeine. The seed contains about 12% protein and 2.6% oil. The young pods and leaves are cooked and eaten as a vegetable in the Far East. The beverage is reputed to have a therapeutic effect on certain heart conditions. The leaves and stems show a blood pressure-lowering effect and experimentally have reduced inflammation. The leaf tea is used in folk medicine to treat rheumatism, snakebite, syphilitic sores and bed-wetting in children. It is consumed by or used as a bath for fever patients and applied externally for skin diseases. The leaves may be spread under a sheet or mashed and applied to the body to relieve fever. The crushed leaves are used as a poultice on tumors and to treat ringworm. The root tea is used to treat a variety of problems with the female reproductive system, rheumatism, snakebite, hysteria and fevers. The seed tea is used to treat several heart problems, cataracts, insect bites and snakebite and to relieve colds. The seeds—and to a lesser extent the whole plant—are purgative.

Notes The generic name is derived from an ancient Greek name for a spice. The species epithet is Latin and refers to the origin of the plant in the Western Hemisphere. Common names for this plant also include wild coffee, stinking senna, glaucous senna and many similar variations. The seeds contain cassiollin, chrysophenol, emodin, physcion and the three trihydroxyquinone derivatives helminthosporin, islandicin and xanthorin. The seed oil contains primarily palmitic (66%) and linoleic (23%) acids. The leaves contain chrysophanol, bianthraquinone and two flavonoid glycosides. The roots contain physcion and phytosterol.

Siamese senna   p. 49
Cassia siamea   Leguminosae

Toxic properties The wood has pockets of a yellow powder similar to chrysarobin called chrysophanhydroanthron. The leaves, pods and seeds contain the alkaloid siamine and several similar ones which are fatal to swine in small doses. Other livestock do not seem to be as sensitive to the toxins. The root and bark contain anthraquinones.

Symptoms Swine are particularly attracted to leaves and seeds, which bring about extensive damage to the internal organs, resulting in sickness and death in a short period. The yellow powder in the wood causes irritation and sometimes brown staining on the skin, inflammation of the eye and respiratory difficulties if inhaled.

Treatment If consumed, plant remnants should be removed immediately from the stomach. The yellow powder from the wood should be washed from the skin or rinsed from the eye as soon as possible.

Beneficial uses The light-brown, soft sapwood is not durable, but the hard, streaked, dark-brown heartwood has been used for bridge construction, furniture, turnery and fence posts. The pods and the heartwood have been used for dyes and the bark for tanning. Senna sometimes is used to provide shade and soil enrichment for coffee and cacao plantations. Rows of this tree with their dense crowns make a good windbreak. The flowers are used in curries. Laboratory experiments have shown extracts to be very effective against malaria parasites. Folk medicine uses the heartwood as a laxative and the root to remove intestinal worms.

Notes The generic name is derived from an ancient Greek name for a spice. The species name refers to the origin in Siam (now Thailand). This plant has also been known as *Sciacassia siamea*. Common names include yellow cassia and kassod tree.

Shake-shake
Crotalaria incana

p. 49
Leguminosae

Toxic properties The entire plant contains several pyrrolizidine alkaloids, including the highly toxic retrorsine, which causes liver and kidney neoplasms. The seeds contain the alkaloid integerrimine, which causes liver and lung lesions. A similar species popularly used as a bush tea in Jamaica has caused veno-occlusive disease in experimental animals. The passage of the toxin to infants in the mother's milk may be the source of the veno-occlusive liver disease of infants whose mothers drink the tea.

Symptoms The consumption of this plant has resulted in the death of livestock.

Treatment No treatment is known for the intoxication produced by this plant. It is best to remove it from livestock pastures and avoid consumption of tea from any part of the plant.

Beneficial uses The plant fixes atmospheric nitrogen into a form usable by plants and is valuable as a soil-conditioning green manure. The root tea is used in folk medicine to treat yellow fever and rashes. The rattling pods are used in folk medicine and magic to treat mute or stuttering children "because the plant speaks."

Notes The genus name is from the Greek *krotalon,* meaning "rattle" or "clapper," in reference to the seeds rattling in the ripe, dry pods. The species name is the Latin word meaning "hoary," referring to the woolly covering on the pod and stem. Another common name is velvety rattlebox.

Yellow lupine
Crotalaria retusa

p. 50

Leguminosae

Toxic properties The major toxins of this plant are the pyrrolizidine alkaloids monocrotaline and retrorsine, which make up about 4% of the weight of the seed and also are present in the foliage. The alkaloids retusamine and retusine also are present.

Symptoms Kimberly horse disease, also known as walkabout disease, is caused by this plant. The symptoms of afflicted horses are compulsive walking in a straight line and butting the head against any object encountered. The neurological symptoms coincide with a steep rise in blood ammonia and severe pathological changes in the liver. Acute doses may result in sudden death due to gastric hemorrhage in hogs. Cattle show bloody feces, nasal discharge and yellow mucus membranes before death in 5 to 10 days. Accidentally contaminated sorghum feed with 1 part per thousand of *Crotalaria retusa* seed was fed at a pig farm for 3 weeks. Six weeks later, symptoms became evident and essentially eliminated the entire herd. The symptoms included loss of appetite, reduced weight gain, depression and lethargy. Pneumonia with labored breathing and a nasal discharge was evident in pigs with low dosages. Necropsy showed kidney, lung and liver damage.

Laboratory studies have shown the alkaloids produce proliferation of the bile ducts and endothelial proliferation of the blood vessels of the liver and lungs. Experimentally, one part per 5,000 in the diet always produced a severe fatal disease. One-fifth of that amount produced liver and kidney damage. In chronic livestock poisoning, death may occur as long as 9 months after exposure, with severe symptoms being evident only within the last few weeks. In experiments with chickens, as few as 10 seeds every other day has killed them in 3 to 6 weeks. One part per two thousand in the diet of poultry results in depression and death due to liver disease within 8 weeks. Monocrotaline is absorbed through the intact skin and produces pathological effects.

Treatment The best course of action is diligent prevention of consumption. There is no known way to prevent damage to the liver once a toxic dose of a pyrrolizidine alkaloid has been absorbed from the intestines. Prompt removal of the stomach contents reduces absorption of the alkaloid. External application of folk medicinal preparations should be avoided due to the risk of absorbing harmful doses of monocrotaline.

Beneficial uses Monocrotaline has been found to be active against adenocarcinoma in the laboratory. This plant has been used as a source of dye and has been called an indigo plant. It contains much indican. In folk medicine the root is used to treat colic and the fresh juice or leaf tea is used for fevers and externally applied to treat skin diseases. The young leaves have been cooked as a vegetable and the fiber from the plant is used to make canvas, cords, nets and ropes.

Notes The generic name is from the Greek word meaning "to rattle," in reference to the seeds rattling in the ripe, dry pods. The species name is the Latin word meaning "notched," referring to the indented leaf apex. Another common name is large yellow rattlebox.

Showy crotalaria
Crotalaria spectabilis

p. 51

Leguminosae

Toxic properties The pyrrolizidine alkaloid monocrotaline, with lesser amounts of spectabiline, found in the seeds, is the primary source of toxicity. The toxin is cumulative. In humans, it produces damage in the form of thromboses of the liver veins leading to cirrhosis. This phenomenon is known as veno-occlusive disease and has repeatedly occurred in the West Indian children given a " bush tea" made from the leaves of this plant. Significant livestock and poultry losses have resulted from feeding them this plant. Experimentally, chickens fed 20 seeds survived with no symptoms, 40 seeds produced symptoms in half the birds, and larger doses induced various levels of chronic or acute symptoms. Consumption of 5 seeds a day produced death. Chickens given adequate diets of mash and scratch feed will pick toxic amounts of the seeds from a field. Quail force-fed 20 seeds did not show symptoms of poisoning, but doses of 40 or 80 seeds resulted in death. When given access to *Crotalaria* seeds under natural conditions, quail do not eat them. Turkeys force-fed over 300 seeds did not show signs of poisoning and when allowed to roam in a field with abundant *Crotalaria* seeds did not eat the seeds. A dove force-fed 80 seeds in a single feeding and another fed ten seeds a day died 19 and 10 days later respectively.

Symptoms Consumption of the seeds induces abdominal pain, vomiting and diarrhea, and loss of appetite. The liver and spleen are often damaged and enlarged. Poultry with acute poisoning show a watery discharge from the mouth and nostrils, a congested comb, ruffled feathers with a droopy appearance, diarrhea, and a full crop. A bird with chronic poisoning gradually recovers from the former symptoms, the comb becomes scaly and loses its color, and the bird appears in poor condition. Necropsy of poisoned fowl shows lesions in the liver and lung. Experimental feeding to swine produces a severe anemia and sometimes lethal gastric hemorrhage. Swine with black markings exposed to low doses of the seeds may lose the hair on the black areas of the body but not the white. Pigs show poor weight gain delayed for weeks after exposure. In cattle, chronic consumption produces a gradual onset of symptoms 1 to 6 months after exposure. Loss of appetite, excessive salivation, yellowish discoloration of the eyes, diarrhea and partial eversion of the rectum are typical. Rabbits may die several months after ingestion of the seeds.

Treatment The best treatment is prevention by vigilance. Most poisonings result from the plant contaminating hay or the seeds occurring in grain-based feeds. In livestock, copper sulfate and ferric ammonium citrate have been reported to bring about dramatic recovery.

Beneficial uses Although this species was introduced and distributed intentionally by the U.S. Bureau of Plant Industry in 1921 as a soil conditioner and forage crop, it was 10 years before the toxic properties of the seeds were discovered. The plant yields abundant nectar which bees make into a dark, strong-flavored honey.

Notes The genus name is from the Greek *krotalon* meaning "rattle" in reference to the seeds rattling in the ripe dry pods. The species name is from the Latin word meaning "spectacular." This plant has also been called *Crotalaria sericea*.

Blue rattleweed
Crotalaria verrucosa

p. 51
Leguminsoae

Toxic properties The primary toxin in the seeds is the alkaloid crotalaburnine (also called anacrotine). Also present is crotaverrine and other related alkaloids.

Symptoms It has been suggested that the use of this plant in traditional herbal medicine in Sri Lanka and other areas is responsible for the high incidence of chronic liver disease and cancer. Livestock eat all stages of the plant and show symptoms as described for the other *Croatalaria*.

Treatment Folk medicinal use of this plant should be avoided and livestock should be prevented from consuming it. There is no known treatment for the toxic effects of this plant.

Beneficial uses The leaves are applied externally in folk medicine to treat headache and used dangerously as an ingredient in herbal teas.

Notes The generic name is from the Greek *krotalon,* meaning "rattle," in reference to the seeds rattling in the ripe, dry pods. The species name is from the Latin word meaning "warty." The common name, purple rattlebox, is also used for this plant. Other common names include purple sesbane and purple rattlebox.

Dwarf poinciana

Daubentonia punicea

p. 52

Leguminosae

Toxic properties The seeds and flowers contain a toxic saponin and 3 closely related glutaramide compounds called sesbanamide A, sesbanamide B and sesbanamide C.

Symptoms After a latent period of several hours to a day after eating the seeds, humans show a loss of appetite, vomiting, abdominal pain and persistent diarrhea (sometimes bloody). If the saponin is absorbed into the bloodstream, depression, muscular weakness, shallow, rapid respiration and a fast, irregular pulse may appear. The saponin produces degeneration of blood cells, kidneys and liver. A death has been reported due to acute liver involvement. After consuming the seeds, livestock show diarrhea, labored breathing, rapid pulse and depression. A lethal dose for sheep has been 50 g (2 oz) of seeds, for chickens 12 seeds and for pigeons 4 seeds. Goats and cattle have also been affected.

Treatment There is no specific treatment for the toxin. If vomiting has not removed the stomach contents, it should be induced and followed by a slurry of activated charcoal. Demulcents such as milk, egg white and vegetable oil should be administered to soothe and protect the intestinal tract. Severe dehydration is possible and may require prompt fluid and electrolyte replacement. Supportive and symptomatic therapy is appropriate.

Beneficial uses This plant is grown as a hardy ornamental. The sesbanamide compounds have shown significant leukemia inhibitory activity.

Notes The genus is named in honor of Louis Daubenton (1716–1799), a naturalist. The species epithet is Latin for "reddish-purple." This plant is also known as *Daubentonia longifolia* and *Sesbania punicea*.

Coral bean

Erythrina corallodendrum

p. 53

Leguminosae

Toxic properties Members of this genus contain toxic curare-like alkaloids which cause paralysis by blocking nerve transmissions to muscles. Some of the alkaloids exhibit the curariform action even when taken by mouth. The seeds of *Erythrina herbacea* are known to contain the alkaloids erysopine, erysothiopine, erysodine, erysovine and hypaphorine. *Erythrina variegata* contains the alkaloid erythrinine, which is a central nervous system depressant.

Symptoms Vomiting, malaise, lethargy and depression have been reported after consumption of the seeds.

199

Treatment The seeds or fragments should be removed from the stomach as soon as possible. Symptomatic and supportive therapy should be applied as needed.

Beneficial uses The showy seeds have been drilled and used in various jewelry and craft items. It is reported that the young leaves and stems have been cooked and eaten, but this is probably a dangerous practice. The seeds have been used to poison rats and dogs. Folk medicine has used teas prepared from the leaves and/or bark to treat asthma, venereal disease, insomnia and hysteria, sometimes with fatal results due to overdose.

Notes The generic name is from the Greek word *erythros,* referring to the red flower color. The species epithet is Greek for "coral tree." Other common names include Cherokee bean and immortelle.

Madre de cacao   p. 53
Gliricidia sepium   Leguminsoae

Toxic properties The leaves are poisonous to dogs and horses, but make nutritious fodder for cattle and goats. It seems likely that the microorganisms in the rumen chemically neutralize the toxic component as they do in the case of *Leucaena.* An unconfirmed claim is that although the roots, seeds and leaves are toxic to mice, they are not toxic to rats.

Symptoms Consumption of a single leaf is reputed to leave a horse ill and incapable of work for more than a week.

Treatment It would be prudent to prevent nonruminant livestock from feeding on the seeds or foliage until further research is completed on the toxins in the plant.

Beneficial uses This tree has been widely planted as an ornamental. The fresh or dried foliage has been used effectively as a protein fodder supplement for cattle and goats. The leaf meal has been substituted for alfalfa in chicken feed to provide a rich color to the egg yolks. The Aztecs found that cocoa grown in the shade of these trees does well. In later years, it has been found that the plant's roots have nodules which fix nitrogen and enrich the soil, while the poisonous seeds, leaves and roots help control rats which otherwise damage the cocoa pods. Interplanted with coconuts, this tree significantly increases yields due to its nitrogen-fixing capabilities. The tree has been used to shade coffee, tea and cloves. It is used as a living fence and as a substrate for growing vanilla orchids and black pepper. The hard, heavy, strong, brown wood is difficult to work due to an irregular grain, but finishes easily and is dimensionally stable. It develops an attractive reddish tint when exposed for some time to the sun. The wood is resistant to termites and marine borers. This species is often used as a street tree, an ornamental or a hedge. The leaf tea has been used in folk medicine to treat colds, fevers, gonorrhea, jaundice, edema and kidney problems. The poulticed leaves are applied to bruises, itches, sores, skin diseases and wounds, and are used to treat headaches and stomachaches. The flowers are sometimes eaten after being boiled in a soup or fried. The seed oil contains 5 times the beta carotene of corn oil and has been recommended for commercial use. Bees collect the abundant nectar for honey.

Notes The generic name is a Latinization of *Glis*, "dormouse," and *cidium,* "kill," thus freely translated as "mouse killer." The species is from the Latin word meaning "hedge." The tree is also known among Spanish-speakers as *mata-raton* for its use as a rodent poison and *madera negra* for its dark wood. *Cacahuanantl*, the original Aztec name for the tree, translates to "mother of cocoa" or the Spanish *madre de cacao*. These names are derived from its use in providing soil improvement and shade in cocoa plantations. *Gliricidia* has been found to be allelopathic in secreting compounds which inhibit growth of other plants.

Poison indigo
p. 54

Indigofera suffruticosa   Leguminosae

Toxic properties The characteristic product of this genus is the dried extract of the plant, called indigo. The purple color of indigo and the toxic properties are primarily due to the presence of the glucoside indigotin. A toxic glucopyranoside, 3-nitropropanoic acid, has been isolated from the roots and stem of this plant. Experimentally, extracts from the stems show toxicity and cause central nervous system depression. The roots have insecticidal properties.

Symptoms Large doses of the tea have caused nausea and vigorous purging. Livestock have died as a result of being fed this plant.

Treatment The toxins associated with this plant are both diuretic and purgative, and as such are soon removed naturally from the body.

Beneficial uses This plant has been used in agriculture as a cover crop and as a green manure. This plant and several similar species have been used as a source of indigo dye for over 4,000 years. The dye is extracted from the leaves and branches by soaking in water made alkaline by the addition of lime. The dyed cloth turns blue after soaking in the indigo solution and subsequent exposure to air. The plant has been used in folk medicine to treat boils, burns, convulsions, diarrhea, dizziness, epilepsy, enlargement of the liver and spleen, fevers, heat rash, inflamed swellings, insect bites and stings, laryngitis, mumps, rabies, scabies, snakebite, syphilis, ulcers, urinary diseases, worms, yaws and very effectively as a purgative. This plant is used frequently in traditional Chinese medicine to reduce inflammation and relieve pain. The powdered roots and seeds or a tea made from them is used extetnally to eliminate ectoparasites and to treat heat rash. The wood has shown experimental antibiotic and androgenic activity. Ecologically, it is used as a cover crop which enhances soil fertility and is said to reduce the numbers of parasitic nematodes. The roots and leaves have been crushed and used as a fish poison.

Notes The genus is derived from the Latin words meaning "indigo bearing." The species epithet refers to the somewhat shrubby growth habit of this plant. This plant is also known as wild indigo. Indigo has been the source of considerable historical intrigue. The importation of indigo threatened the European producers of woad (a plant producing a similar blue dye), so they arranged to prohibit the importation of indigo. Eventually, indigo was legally allowed into Europe. The West Indies surpassed India in indigo production for the European market in the mid-18th century. High duties eventually returned indigo production to India, but the planters had such unpleasant working arrangements in Bengal that a labor revolution equivalent to emancipation occurred. Indigo essentially was replaced by aniline dyes by the end of the 19th century. Chemical examination of extracts has shown the presence of the prenylflavone louisfieserone, beta sitosterol and pinitol.

Wild tamarind
Leucaena leucocephala

p. 55
Leguminosae

Toxic properties The leaves and mature seeds contain the toxic amino acid mimosine, which inhibits DNA synthesis and thus has particular influence on the rapidly dividing cells producing hair growth. The mimosine content is highest in rapidly growing leaf tips. In ruminant animals, the mimosine is converted to 3,4-dihydroxypyridene, which impairs the incorporation of iodine into the thyroid, thus producing goiter-like symptoms. The toxic properties are mostly inactivated by cooking the plant in an iron or aluminum pot or roasting the seeds.

Symptoms Excessive consumption of the seeds, pods or leaves causes hair loss in humans. The hair loss is accompanied by pain and swelling of the scalp and eyebrows. This compound causes loss of hair in horses, mules, donkeys and pigs, and in certain populations of cows and goats. Mimosine and/or its metabolites induces hypothyroidism, with subsequent goiter symptoms in cattle and sheep, along with a reduction in growth and development of more body fat. Cataracts are more abundant in animals consuming a diet high in *Leucaena*. Embryonic death and resorption of the fetus is common in livestock with *Leucaena* in the diet. *Leucaena* and mimosine are teratogenic and cause a variety of fetal malformations. Mimosine exerts an allelopathic effect by inhibiting the growth of plants when present in the soil.

Treatment Consumption of this plant by nonruminant livestock should be limited to prevent adverse reactions. It has been found that a gram-negative anaerobic bacteria living in the rumen of cattle and goats in Hawaii will metabolize mimosine to a nontoxic compound. This bacteria, introduced to the rumens of livestock in Australia, has allowed them to feed on *Leucaena* with impunity. The addition of ferrous sulfate to the diet will allow mimosine to form complex iron compounds which are excreted in the feces. Cooking with moist heat will reduce the toxicity of the plant when fed to livestock.

Beneficial uses Because its roots are capable of fixing nitrogen from the air, this plant can grow well on very poor soil and is drought-tolerant. It grows rapidly and is an excellent source of firewood in less-developed countries. When it is allowed to become a tree, the lumber is of a golden-brown color and is suitable for flooring. When reduced to pulp, it produces a high-quality paper. The bark produces a brown dye. When cut, it coppices vigorously and produces a foliage with up to 35% protein in the growing leaf tips. It can be cut and dried and used in place of alfalfa in livestock feeds. It makes an excellent fodder for suitable ruminant livestock. Small farmers commonly gather it from roadsides to provide supplemental feed to goats. The depilatory effects of mimosine have been utilized in defleecing sheep by feeding high doses for short periods of time. Each follicle simultaneously produces a weakened hair, allowing the wool to be peeled easily from the sheep. The seeds have lowered the blood sugar and reduced the serum cholesterol of diabetic rats.

Folk medicine has used the leaf tea to treat back pain, menstrual cramps and typhoid fever and to alleviate intestinal gas. The root and bark tea have been used to induce menstruation and abortion and as a contraceptive. Experimentally significant decreases in conception by rats with 15% *Leucaena* leaf in their diets have been recorded. The immature pods and flowers have been cooked and eaten as a vegetable. The seeds have been cooked as a delicacy and roasted as a coffee substitute. The seeds are often incorporated in handicrafts such as hat bands, belts and necklaces. Bees collect the nectar and abundant white pollen.

Notes The genus name is from the Greek word "leukos," referring to the white flowers. The species epithet is Greek, meaning "white head," referring to the spherical white flower heads. Older and now generally abandoned names for this plant include *Leucaena glauca, Mimosa glauca,* and *Acacia glauca.* Of the many other common names of this plant, two which are also in frequent use in the West Indies are tan-tan and lead tree. In Hawaii, it is known as *kao haole* and in Southeast Asia *ipil-ipil.* The seeds contain 8.5% fat. The percentage of the amino acids in leaf protein concentrate is arginine 6.4, histidine 2.2, isoleucine 5.0, leucine 9.1, lysine 6.3, methionine 2.4, phenylalanine 5.9, and threonine and valine 6.0 each.

Jerusalem thorn   p. 56
Parkinsonia aculeata   Leguminsoae

Toxic properties The plant is known to accumulate lethal levels of nitrates. In ruminant animals, nitrates are reduced to the readily absorbed nitrite ion. When the nitrite ion is absorbed into the bloodstream, it converts hemoglobin into methemoglobin, which carries less oxygen to the tissues, leading to hypoxia. Several polyphenolic glycosides have been found in the leaves.

Symptoms The first signs of nitrate poisoning are a shortness of breath, becoming more severe until panting, gasping and severe apprehension are evident. Weak, rapid heartbeat, reduced body temperature, loss of balance, staggering gait, frequent urination and muscular weakness or tremors may be present. In acute cases, convulsions and/or coma may lead to death within 3 hours. Pregnant females may abort after acute or chronic exposure to high levels of nitrates. Effects of chronic exposure are decreased milk yield, abortion and disturbed vitamin A metabolism.

Treatment For livestock, an intravenous injection of 20 ml of methylene blue (a solution of 10 g in 500 ml of water) per 100 lb of body weight aids in the reduction of methemoglobin to hemoglobin.

Beneficial uses The reddish-brown, hard, heavy heartwood is seldom used except for fuel and crude construction. The wood has been shown to be suitable for paper-making. Folk medicine has used the leaves as an antiseptic, to induce abortion and to treat fevers and tuberculosis. A highly questionable claim has been published in the Indian medical literature, reporting the juice as an effective treatment for rabies. The leaves contain 7.5% protein. The beans have considerable potential as a human food source, with a 21% protein content and low levels of flatulence-producing oligosaccharides. The beans are rich in K, Ca, Mg, and Fe. The dietary benefits may have been recognized by the Native Americans, who ground the seeds to produce a meal or flour. It may have been one of the few plants to reliably produce a nutritious food in semidesert lands in drought years.

Notes The generic name honors John Parkinson, an English pharmacist and author of two important botanical books. The species epithet is Latin, meaning "prickly" or "spined." It is also known by the common name of Mexican palo verde or horse bean. In Indian folk medicine, it is called garur buti.

Cowitch   p. 57
Stizolobium pruriens   Leguminosae

Toxic properties The active ingredients in the barbed, stinging hairs are a proteolytic enzyme called mucanain, which causes the itching, combined with serotonin, which causes the pain. The seeds contain L-DOPA (3,4-dihydroxyphenylalanine). The entire plant contains serotonin (5-hydroxytryptamine), mucuadine, mucuadininene, nicotine and prurienidine. The seeds contain the alkaloids mucanadine, mucunine, prurienine and prurieninine. The latter two compounds have the physiological effect of dilating the blood vessels, lowering the blood pressure and increasing peristalsis.

Symptoms Three centuries ago, it was written that "the slightest particle of the pod covering causes an intolerable itching when rubbed on the skin." The stinging hairs retain their potency for long periods, even when detached from the pod, and have been marketed as "itching powder." The exquisitely effective stimulus of stinging and burning produced by the piercing of the epidermis and the release of the toxin from the hairs may be quite far removed in distance and time from the presence of the pods. The experimental insertion of a single spicule has been shown to produce an intense itching within 30 seconds. A spicule autoclaved at 250°C for 30 minutes or immersed in boiling water is not visibly changed, but produces no symptoms. The itching and stinging rash is particularly prevalent among workers in sugar cane fields which harbor the plant and among bulldozer operators clearing scrub vegetation. The seeds are toxic when consumed in excess by horses, chickens and pigs. They cause vomiting and diarrhea in swine. The toxic principle is reported to be L-DOPA.

Treatment Once symptoms are present, the only treatment is to launder clothing and rinse off any hairs which may still be present on the skin. Toxic symptoms induced by excessive seed consumption resolve themselves after the seeds are eliminated from the diet.

Beneficial uses The hairs have been sold commercially as an itching powder and used as a scientific tool in the study of itching since the 1930s. Folk medicine (and some officially sanctioned medicine such as older editions of the *British Pharmaceutical Codex*) has used the hairs from the seedpod mixed with honey or molasses as a vermifuge for roundworms, threadworms and tapeworms. Experimental microscopic observation of intestinal worms exposed to the spicules showed the worms "began to writhe and twist themselves . . . and exhibited evident signs of great torture." The leaves are used as a tonic, aphrodisiac and anthelmintic, to reduce inflammation and to cure headache. The aphrodisiac action is probably associated with the presence of 5-methoxy-N,N-dimethyltryptamine, a well-known hallucinogenic agent. The seeds contain up to 4% L-DOPA and have been used in Ayurvedic medicine for 3,000 years to treat symptoms of parkinsonism.

Modern medicine now routinely uses L-DOPA to relieve the symptoms of parkinsonism. Experimentally, the seeds have also been found to reduce blood sugar and cholesterol levels in the blood. In folk medicine, the seeds are made into a tea used as an aphrodisiac, astringent, diuretic and laxative, nerve tonic, and to treat various male and female reproductive disorders and scorpion sting. The root is used to treat cholera, gout and erysipelas, as a diuretic and is smoked to reduce pain and accelerate delivery in childbirth. The immature pods are cooked and eaten as a vegetable. Young seeds, which have over 16% protein and 17% carbohydrate, are cooked and eaten by many indigenous people. Usually the seeds are boiled and soaked in several changes of water to remove the excess L-DOPA. The seeds are also fermented after boiling to produce a bean cake similar to the tempeh made from soybeans. The seeds have been used in the manufacture of miso. The high oligosaccharide content may lead to a considerable production of flatulence. Dry roasting of the seeds destroys the oligosaccharides and eliminates the flatulent effect of the seeds. Mature dried seeds have been roasted, ground and used as a coffee extender or substitute. The immature pods and leaves are sometimes cooked and used as a vegetable. The cultivated forms are used as a smother crop to control weedy grasses or as a source of green manure. The plant serves as high-protein forage crop for livestock when planted as a pasture or gathered as hay or silage.

Notes The generic name is from the Greek *stizo,* meaning "to prick," and *lobos,* meaning "lobe" or "pod." The species epithet is the Latin word for "causing an itch." The name *Mucuna pruriens* is also much used for this plant. This plant is also known commonly as cowage, itchweed and pica-pica. Cowitch is a perversion of the Hindu name *kiwach,* meaning "bad rubbing." The stinging hairs applied to the skin followed by the latex of *Euphorbia* produce a persistent swelling and numbness which simulates leprosy. The seed contains up to 12.5% lecithin, making it a candidate for use in confectionery and other industrial applications. The seed oil is composed of 30% palmitic acid, with 15% each linoleic and lenolinic acids and 10% each palmitoleic, oleic, myristic and stearic acids. Other members of this genus climb to the canopy of the rain forest and provide nutritious seeds for many animals. The alkaloid 6-methoxyharman has been identified from the leaves. Bees produce an abundant golden honey with an inferior taste from the blooms.

Climbing lily
Gloriosa superba

p. 58
Liliaceae

Toxic properties The root, leaf, stem, flower and seeds are all poisonous due to the presence of colchicine. The seeds contain the highest amount of toxin per unit weight, but no record is present of harmful amounts being consumed. The tubers contain approximately 0.3% colchicine and lesser amounts of the similar gloriosine, yielding a potentially lethal dose if 4 g (0.14 oz) of the tuber are consumed. Colchisine is excreted in the milk, and human poisoning has been reported due to the consumption of milk from colchisine-poisoned livestock. Elimination of colchisine from the body is relatively slow; thus, the toxin may accumulate in the body from subtoxic daily doses. Both gloriosine and colchisine have an antimitotic effect and tend to stop cell division by disrupting the spindle apparatus during metaphase. Their greatest effect is on rapidly dividing cells such as those in tumors, bone marrow, intestinal epithelium and hair-producing cells. Colchicine increases sensitivity to central depressants and sympathomimetic agents, lowers body temperature, constricts the blood vessels and raises the blood pressure, enhances digestive tract activity by nervous stimulation and alters neuromuscular function. Colchicine is able to withstand drying, storage and boiling; thus, the cooked tubers and dried seeds retain full toxicity.

The alkaloid lumicolchicine is found in the flowers. Other publications report the presence of a toxic compound, superbine, which is experimentally lethal to a cat in a dose as small as 0.01 g (0.00036 oz). This compound is suggested to be very similar to the toxic, bitter compound in squill, or sea onion, which is used in several proprietary rodent poisons.

Symptoms A tingling, burning and subsequent numbness is felt in the lips, mouth and throat immediately or—more usually—after 3 to 6 hours. Symptoms may be delayed for as long as 48 hours after ingestion. Peak plasma levels of colchisine develop within 30 to 120 minutes of ingestion. Difficulty in swallowing and a feeling of strangulation may be present. Intense thirst and sometimes violent vomiting are followed by abdominal pain and diarrhea which may be so severe as to become bloody and cause dehydration. From one to several days after ingestion, skin numbness of the extremities, weakness in the extremities, discomfort from bright light, low blood pressure with a rapid weak pulse (slowed pulse has also been reported) and decreased or absent urine production may give way to gradual ascending paralysis of the central nervous system and loss of deep tendon reflexes, leading to death from respiratory failure. Control of body temperature may become erratic, resulting in subnormal temperatures 20 to 40 hours preceding death. Convulsions leading to unconsciousness, or shock due to fluid volume depletion, may occur. Death has occurred during a relapse after an apparently symptom-free period. Bone marrow depression follows ingestion within 4 or 5 days. Complete hair loss may begin within 12 days and be complete within 3 weeks. Bone marrow function, hair growth and deep tendon reflexes resume in survivors after a month.

Treatment   Prevention includes storage of the tubers in labeled containers away from food storage areas to prevent mistaken cooking and consumption. There is no specific antidote, but tannic acid has been recommended to precipitate the colchicine in the intestinal tract and prevent further absorption. In any case, the residual plant material should be cleared from the gastrointestinal tract if natural processes have not accomplished this before medical aid is sought. A neurogenic source is suspected for the diarrhea, as experiments have been able to control it with atropine and anesthesia. Symptomatic and supportive care, including fluid and electrolyte replacement and monitoring of blood pressure, respiratory and kidney function, are appropriate. Meperidine alone or combined with atropine may be needed to counteract the severe abdominal pain. Respiratory assistance may be needed in the case of muscular paralysis. One patient who consumed 125 g (4.5 oz) of the tubers survived after medical treatment with vasopressors, fluid replacement and steroids. There is no specific effective antidote for colchicine. It is primarily broken down by the liver, but up to 20% may be excreted unchanged by the liver. Large amounts of colchicine and its metabolites are excreted in the bile and other intestinal secretions. Other reports show 30% of colchicine excreted in the urine within 24 hours and up to 97% within 48 hours.

Beneficial uses   The plant is a popular and attractive ornamental. Preparations of colchicine have been used effectively in the treatment of acute gout since 1763. The mechanism of action is unknown. Cochicine has also been suggested as a remedy in the treatment of arthritis, cholera, colic, skin diseases, kidney problems and typhus. It has been widely used as an experimental tool in the study of cell division. The plant is used in folk medicine, but should be avoided or approached with extreme caution, as overdoses have caused deaths. The root has been used to induce labor and abortions, as a tonic and to treat bleeding from the nose, bruises, cancer, gonorrhea, hemorrhoids, leprosy, malaria, nocturnal seminal emissions, impotence, scabies, sprains and syphilis. It is also used around doors and windows to repel snakes and, if that fails, it is used to treat snakebite or scorpion sting. The juice of ground *Gloriosa* leaves has been used to control head lice.

Notes   The generic name is Latin, meaning "full of glory," and the species epithet makes reference to the superb flower. This plant is also called glory lily and gloriosa lily. The root has caused many accidental deaths and has been used to commit homicide, suicide, to induce abortions and to poison dogs. Colchicine is eliminated from the body by excretion into the intestinal tract, but while in the blood it is excreted in the milk of humans and other mammals. A lethal dose of colchicine in an adult has been as little as 3 mg (0.0001 oz). Colchicine in plant cells interferes with spindle formation, resulting in doubling of the number of chromosomes, a phenomenon called polyploidy. Polyploidy is passed on to subsequent generations of plants in the seed and has been used successfully by plant breeders to produce new cultivars. The flowers contain the related compounds beta-lumicholchisine, 3-desmethylcochicine and N-formyldesacetyl-colchisine. Other alkaloids present in the tuber are N-formaldesacetylcolchicine, dimethylcolchisine, gloriosine, gloriosol and lumicholchisine. Salicylic acid and resins are also present in the plant tissues. African porcupines and Florida moles are reputed to be able to consume the roots with impunity.

Carolina jessamine

p. 59

Gelsemium semperviren

Loganiaceae

Toxic properties All parts of the plant contain toxic alkaloids. The most abundant are gelsemine, gelsevirine, gelseminine, gelsemicine, gelsemoidine, gelsenicine and koumidine. The alkaloids are most abundant in the roots and have a cumulative effect on the nervous system. Gelsemine is the most abundant alkaloid in the flowers and has been shown to adversely influence nerve transmission by serving as a potent antagonist of muscarinic and nicotinic acetycholine receptors. The nectar is made into a toxic honey which is suspected of having caused several human deaths.

Symptoms In humans, symptoms have become evident within 10 to 20 minutes of consuming a decoction of the roots. These include pain in and above the eyeballs, double vision, dilated pupils, dizziness, nausea, muscular weakness, staggering gait, weak pulse, depressed temperature, sweating and slow respiration, which may lead to suffocation. Children have been intoxicated by sucking the nectar from the flowers. Flocks of poultry, particularly turkeys, have been devastated by habitually scratching up and consuming the roots of this plant, which produces the disease called limber neck. Pigs, goats, sheep, cows and horses have been lethally poisoned by chronic consumption of this plant. Symptoms of acute poisoning in livestock include a weak, feeble pulse, slow breathing, depressed body temperature, dilated pupils and muscular weakness. Eating the leaves has caused abortion in cattle. Individuals handling the roots are subject to a dermatitis from the contact.

Treatment The stomach contents should be removed immediately and replaced with a slurry of activated charcoal. Fluid replacement and respiratory support may be needed in extreme cases. One case responded well to intravenous fluids at 1.5 times maintenance amounts, acidified with ascorbic acid to produce a urine pH less than 6.0. Morphine has been recommended as a specific antidote. Atropine has also been suggested.

Beneficial uses This luxuriantly flowering vine is grown as an ornamental at low elevations from Virginia to Texas. Sempervirine and other alkaloid compounds in methanol extracts from the stem have been found to be active against cancer cells in laboratory tests and have cured mice of several types of cancer. Gelsemine is analgesic, slows the heartbeat and reduces blood pressure. The *British Pharmaceutical Codex* has noted the use of the powdered roots to treat trigeminal neuralgia and migraine headache, but cautions that excessive doses cause double vision, giddiness and respiratory depression. Various formulations derived from the root have been used in folk medicine to treat asthma, diarrhea, dysentery, elevated blood pressure, insomnia, tetanus, toothache, rheumatism and externally for pains such as facial neuralgia. Fatalities have resulted from overdose when used medicinally. The medicine produced by steeping roots in whiskey has been particularly subject to abuse and has been a cause of death.

Notes The generic name is a Latinization of *gelsomino*, an Italian name for jasmine. The species epithet is the Latin word meaning "evergreen." This is the state flower of North Carolina. The honey from this plant is reputed to be toxic and has been held responsible for human deaths.

Mistletoe   p. 59
Phoradendron piperoides   Loranthaceae

Toxic properties The major toxic compound in the leaves and stems is the toxalbumin phoratoxin, which inhibits protein synthesis in the intestinal wall. The berries contain toxic proteins and the toxic amines tyramine and beta-phenylethylamine. The alkaloid rubrine C, extracted from *Phoradendron rubrum*, produces a fall in blood pressure and sustained uterine contractions.

Symptoms Children are usually exposed after eating the berries. Adults more frequently show symptoms after drinking a tea to induce abortion or using mistletoe as a folk medicine. Vomiting, stomach cramps, diarrhea and copious urination follow within an hour or so of consumption. In large doses, dilated pupils, mental confusion, seizures, shock and death due to cardiac collapse may follow. A death has resulted from drinking a tea made from the berries. Injections of phoratoxin produces low blood pressure with slow, weak heartbeat and constriction of the blood vessels. Experimentally, an extract of *Phoradendron* has been found extremely toxic to mice. A dermatitis has been reported from contact with the plant. Hepatitis has resulted from chronic ingestion of an herbal remedy containing mistletoe.

Treatment Residual plant material should be removed from the stomach and replaced with a slurry of activated charcoal. Fluid and electrolyte replacement may be needed in extreme cases. Seizures may be treated with intravenous diazepam.

Beneficial uses The Greeks wrote about other mistletoes 4,000 years ago. They have been used medicinally by native people in America, Japan, Malaya, Australia, India and Europe. Folk medicine has used mistletoe in various forms to treat gout, heart disease, syphilis and epilepsy, to reduce high blood pressure, to induce abortion, and as an antispasmodic and a calming agent. Caribbean folk medicine has used the tea as a remedy for colds, whooping cough, intestinal gas, and as a bath to cure children of marasmus. Native Americans have used the tea to relieve toothache and induce abortion. The plant is reputed to have an action similar to that of oxytocin in stopping postpartum hemorrhage. Many folk remedies require mistletoe growing on certain host plants; indeed, it has been found that mistletoe may accumulate chemicals unique to the host tree—for instance, caffeine is present in mistletoe growing on coffee trees. An extract of the plant kills mosquito larvae. Birds relish the berries and disperse the plant by leaving seeds in their droppings on new hosts. The plant is generally considered to be an obnoxious parasite, but certain small businesses thrive on an annual harvest prior to the Christmas season. Horticulturists sometimes grow mistletoes as botanical curiosities. The plants have been cut from their host trees as fodder for livestock during times of drought, but this practice is considered dangerous, as it is likely to induce abortion in pregnant cows. This is an important honey plant when abundant, producing a very sweet product which is gluey and hard to extract from the comb.

Notes The generic name is from the Greek words meaning "thief" and "tree," in reference to its parasitism on trees. The common name is from the medieval Anglo-Saxon *mistltan,* meaning dung-twig, referring to the folkloric belief that the plant could reproduce only when the seed was deposited in bird droppings. The early Celts used mistletoe in their ritual ceremonies associated with the winter solstice and renewal of life. As Christianity superseded earlier religions, mistletoe was adopted into social tradition at Christmas, held at the same time of year. Americans have further modified the tradition by hanging the branches of the plant in their houses at Christmas in the belief that it has a supernatural power of inducing kissing.

Wild cotton   p. 60
Gossypium hirsutum   Malvaceae

Toxic properties The pigment glands in the seeds, leaves, stem and root bark contain the toxic dihydroxyphenol gossypol, which is insoluble in water but soluble in oil and hydrocarbon solvents. One of the effects of gossypol is to chelate iron in the liver and interfere with the synthesis of hemoglobin and respiratory enzymes. Gossypol also reacts with the amino groups of amino acids and proteins, lowering the nutritional value of the diet and interfering with protein synthesis. Gossypol also adversely affects many physiologically active enzymes. Other terpenoids present include hemigossypolone and various forms of heliocide, which are insecticidal.

Symptoms The typical source of poisoning is consumption of cotton seed, inadequately detoxified seed meal, or seedcake containing gossypol . In adult ruminant livestock, small amounts of gossypol in the diet are tolerated, as the toxin is bound to protein while in the rumen. High doses of gossypol in ruminant animals or lower doses in nonruminants may produce respiratory difficulty, cyanosis and convulsions, which may lead to death within 30 minutes. Hogs are particularly sensitive to the toxin and may show serious swelling and congestion of the liver when the diet contains 2 parts per 10,000 of gossypol. Sows rooting in harvested cotton fields often abort. Calves and lambs become unthrifty, develop anemia, and die after chronic exposure. Acute exposure produces blood in the urine, bloody diarrhea, jaundice and muscular weakness. Respiratory difficulties, abortion and blindness sometimes occur. Postmortem examination reveals general hemorrhage of viscera with lung congestion and heart dilation. Cottonseed cake in the diet of milk cows may result in a poor-quality, gummy butter which takes a long time to form in the churn.

Treatment Iron salts have been reported to decrease the effects of gossypol toxicity in cattle, and the adverse effects of ingestion of the seed by chickens can be somewhat offset by the feeding of iron salts. Gossypol is gradually excreted in the feces by way of the bound compound in the bile, but the addition of ferrous iron to the diet greatly increases the rate at which both free and bound gossypol are excreted. Ferrous iron also detoxifies gossypol by catalyzing its decarbonylation.

Beneficial uses The cotton plant is the source of the world's largest textile industry, with additional thousands of subsidiary uses for the fiber surrounding the seeds. The seeds are a byproduct of the fiber production, but have developed a market of their own for the extracted oil and the residual cottonseed meal. The latter must be carefully detoxified by cooking before it can be used as animal feed. The seedcake may produce intestinal obstructions if fed in excess, due to an indigestible fiber content of as much as 25%. In horses and mules, the high protein content may cause adverse side effects if fed at the rate of more than 450 g (1 lb) per day. The high protein content yields a valuable manure with a high nitrogen content. Cottonseed oil is used pharmaceutically as a solvent for steroid hormone and other injections, as a liniment and as an emollient cathartic. It is used industrially in the manufacture of soaps and cosmetics; as a substitute for sesame oil, a commercial source of vitamin E; and as a solvent, lubricant or ingredient for thousands of other products. Investigation of childless couples consuming poorly refined cottonseed oil led to the discovery that gossypol is a male contraceptive. A study using 10,000 volunteers has shown that 20 mg of gossypol per day produces infertility. Fertility is regained after discontinuation of the drug. Gossypol now is being used as an oral male contraceptive in China. Compounds containing it have been patented in the U.S., but it has not yet been accepted in Western medicine. Gossypol has been demonstrated to be spermicidal, and tests on primates indicate promise for its use as a vaginal contraceptive. A contraceptive product with Gossypol as the primary active ingredient has been patented.

Gossypol has been shown to produce a significant decrease in serum cholesterol, LDL and VLDL when included in the diet. It has also shown anticancer activity in several new cancer drug-screening tests in laboratory animals. Experimentally, gossypol has completely immobilized *Trypanosoma cruzi*, the cause of Chagas' disease, and shows promise as a therapeutic agent for this and other trypanosome-caused diseases. Mice treated with gossypol show almost complete immunity to influenza virus as well as resistance to other viruses, including *Herpes simplex*. Gossypol has also shown antifungal, antibacterial and antitumor activity. Folk medicine has used the leaf tea to treat colds, diarrhea, dysentery, headache, hypertension, respiratory inflammations, rheumatism, stomachache and urinary problems. The root has been used by herbalists as a uterine stimulant to promote abortion, menstruation and parturition, to induce lactation and to treat uterine hemorrhages. The root bark was at one time an official drug in the U.S. Pharmacopoeia and is still popular in homeopathic medicine. It acts as a vasoconstrictor, and promotes menstruation and stimulates delivery by promoting uterine contractions.

Notes The genus is from the Greek and Latin word meaning "cotton." It ultimately is derived from the Arabic *al qutn*. The species epithet is from the Latin word meaning "hirsute (hairy)." The common name in Spanish is *algodon*, and in French *coton*. It was widely used by Native Americans prior to the arrival of Europeans. Because of commercial interests, many varieties and hybrids of cotton have been developed and transported around the world, often escaping and persisting after cultivation. The seed contains up to 33% protein, 16% fat and significant amounts of beta-carotene, thiamine, riboflavin and niacin. The seed oil is composed primarily of linoleic, oleic and palmitic acids, with lesser amounts of stearic, arachic and myristic acids.

Chinaberry

p. 61

Melia azedarach

Meliaceae

Toxic properties All parts of the plant contain varying proportions of a saponin, tetranortriterpenes, neurotoxins, the narcotic alkaloid azaridine (also called mangrovin) and the bitter alkaloids margosine (azadarin), tazetine and paraisine (ocaziridine). The primary toxin responsible for the symptoms has not been clearly identified. It is reported that some cultivars have edible fruit, but the fruit is usually the most poisonous part of the plant, with the ripe fruit more toxic than the green. The toxicity of the plant varies seasonally and from year to year.

Symptoms The fruits have caused deaths in humans, poultry, pigs, cattle, sheep, goats, rabbits and donkeys. Depending on the relative proportions of the toxins in the fruit, the symptoms, which are often delayed several hours, are variable and may include cold limbs, grinding of the teeth, diarrhea, extreme thirst, faintness, loss of equilibrium, dilated pupils, mental confusion, sweating, vomiting and rapid, weak pulse. Severe intoxications may lead to paralysis, depressed irregular respiration and subsequent death by asphyxia within 12 to 15 hours. Liver, kidney, lung, brain and gastrointestinal damage have been reported as a result of consuming the bark tea. The leaf tea has produced bloody vomiting and reduction in the urine flow, along with other symptoms similar to an overdose of belladonna. Hind limb paralysis is common with hoofed animals.

Of all animals tested, swine are the most susceptible to the poison, with 200 g (1/2 lb) of the fruits producing symptoms within 30 minutes, and death by convulsions and respiratory failure within 2-1/2 hours in a 35-kg hog. In swine, the brain, heart, lungs, liver, kidneys and gastrointestinal tract all show hemorrhage. Autopsy of other livestock shows that deaths within a day or so due to acute poisoning show only gastrointestinal irritation, but those that survive several days show degeneration of the liver and kidneys. In laboratory experiments with rats, a necrosis of the skeletal muscles and hind limb paralysis resulted from chronic consumption of a diet containing 25% leaves. Wild birds are sometimes rendered unconscious for some hours after feeding on the fruits.

Treatment The remaining vegetable matter should be removed from the stomach and gastrointestinal tract as soon as possible if vomiting and diarrhea have not done so. Individualized care should be provided for the symptoms demonstrated. Typically, fluid and electrolyte replacement may be needed, along with monitoring of respiratory, liver and kidney functions. There are unsubstantiated folk medicine claims that sugary drinks are an effective remedy for children who have been poisoned by the fruit.

Beneficial uses The attractive reddish-brown wood takes a good polish and is termite-resistant but is soft and brittle. It has been used for fuel, cabinets, furniture, cigar boxes and low-quality construction timbers. Native peoples in the Far East add the leaves to other greens to be boiled as a vegetable. Experimentally, a water extract of green leaves reduced the fertility of male mice; they regained it within 45 days of the discontinuance of the treatment. An aqueous extract of the heartwood is reputed to relieve asthma. The fruit, flowers, leaves, bark and roots have been used in various folk medicine formulations, especially by Arabs and Indians, to expel intestinal worms and as a cathartic. The root bark is official in the Chinese pharmacopoeia. Chinese traditional medicine uses the bark and the fruit to treat various internal ailments and eliminate intestinal worms, as a laxative, and externally to treat ringworm and weeping lesions. It has been used both externally and internally in folk medicine to treat asthma, eczema, fevers, leprosy, ringworm and other skin diseases, to relieve rheumatism and to stop diarrhea.

Numerous human deaths have been reported as results of misuse of parts of this plant in folk medicine. The leaves, stems and bark are insecticidal to crickets, cockroaches and the mosquito *Aedes aegypti*. The leaves and fruit have been found to contain the triterpenoid azadirachtin, which inhibits feeding and growth in the desert locust *Schistocerca* and other insects. Other authors report that an aqueous extract of the leaves containing the carotenoid-like substance meliatin has proven effective in protecting gardens and orchards from attack by locusts and other insects. Meliatin is not toxic, but rather renders the foliage unpalatable. A 2% extract of the mature fruits has been found to be effective in controlling populations of bugs in crop plants. An extract from the fresh leaves has been found to prevent the replication of swine pseudorabies (herpes virus) in laboratory cultures. The leaves and fruit are reported to be insecticidal and have been used to protect stored books and clothing from insect damage. Additional investigation may determine if there are several compounds or various interpretations of the same one. A flammable oil suitable for illumination and lubrication has been extracted from the berries. The seeds are widely used as beads. The bark and crushed leaves have been used as a fish poison. Its early blooming makes it valuable as a honey-producing plant.

Notes The genus is from an ancient Greek name for the ash, whose foliage is superficially similar. The species name is a contracted form of a Persian vernacular name meaning "noble tree." A cultivar of this tree with radiating branches and drooping foliage is known as the Texas umbrella tree, and another form with masses of flowers is known as floribunda. The tree is also known as paraiso, lilac, Persian lilac, lilaila and many variations of the word lilac as well as pride of India, bead tree, paradise tree and hog bush. The use of the seeds as rosary beads is the source of the common name paternosterbaum sometimes applied to this tree. The tree is considered sacred in the

Middle East, but in other areas it is considered a persistent, costly nuisance. The seeds yield a high percentage of oil composed of stearic, palmitic, valerianic and butyric acids.

Velvetleaf   p. 62
Cissampelos pareira   Menispermaceae

Toxic properties The plant and particularly its root contains a great variety of alkaloids comprising 0.72% of its weight. The alkaloids are primarily bisbenzylisoquinolines, with hayatine predominating and making up 0.1% of the weight of the plant. Also present are bebeerine, cissampeline, hyatinine, isochondodendrine and sepeerine.

Symptoms The root tea produces depressant and sedative effects. The plant shows cardiac toxicity and may produce heart failure.

Treatment The plant remains should be removed from the stomach to prevent further absorption of toxins.

Beneficial uses It makes an attractive ornamental plant. Rope and string have been made from the bark fiber. When injected, hayatine methiodide from the root has a neuromuscular blocking action similar to that of tubocurarine, but with greater potency, less toxicity and a greater margin of safety. This drug has been used as a promising muscle-relaxing agent in surgery and has been found to be free of deleterious side effects. The injected drug produces a rise in blood pressure, stimulation of respiration, an increase in somatic reflexes and sometimes convulsions. Cissampareine, one of the root alkaloids, has been shown to have significant antitumor activity. Experimentally, an alcohol extract of the root has shown anticonvulsant properties. The root is used by indigenous peoples to make an intoxicating drink. Folklore has used the root tea to treat the bites of venomous animals and to induce abortion. The tea is diuretic and has also been consumed to treat asthma, bronchitis, burns, boils, cancer, fevers, cholera, colds, cystitis, delirium, diabetes, dysentery, epilepsy, heart trouble, intestinal tuberculosis, hypertension, jaundice, leprosy, malaria, pimples, rabies, rheumatism, snakebite, swellings, and urinary and uterine complaints. The roots are officially listed in the Chinese pharmacopoeia as an analgesic, to reduce swelling, and to treat pain and bleeding of traumatic injuries. The Ayurvedic system of medicine has used the roots as a substitute for tubocurarine.

Notes The genus name is derived from the Greek word meaning "ivy-vine." The species name makes reference to its superficial resemblance to some species of *Pareira*. The plant is also called ice vine and is widely known by the Hindi name of *parha* or *parhi* when it is traded in medicinal commerce. The root alkaloids include bebeerine, cissamine, cissampareine, curine, cyclenine, dehydrodicentrine, dicentrine, hyatine, hyatinine, insularine, isochondodendrine, menismine and pareirine.

Horseradish tree   p.63
Moringa oleifera   Moringaceae

Toxic properties The bark contains sitosterol and the alkaloids moringine and moringinine, which stimulate the heart, relax the bronchioles and smooth muscle fibers, raise blood pressure by peripheral vasoconstriction and act on the nervous system in a manner similar to that of ephedrine. The seeds contain the glycosidic mustard oil 4 (alpha-L-rhamnosyloxy) benzyl isothiocyanate, which does not seem to be present in harmful quantities after cooking. Spirochin, an alkaloid in the root, accelerates and amplifies the heartbeat at 35 mg/kg, but has the opposite effect at doses of 350 mg/kg or larger. It paralyzes the vagus nerve and other parts of the central nervous system and has an analgesic effect.

Symptoms The crushed green leaves irritate and blister the skin externally and act as a purgative and diuretic internally. The leaf tea has a strong purgative action. The root bark in doses of 15 g (0.5 oz) induces abortions.

Treatment The sap and leaves should be washed from the skin as soon as possible after exposure. Internally, the effects seem to be self-limiting.

Beneficial uses The antibiotic compound pterygospermine, extracted from the root bark, is antibiotic to a broad range of gram-negative and gram-positive bacteria, including *Micrococcus pyogenes, Bacillus subtilis, Aerobacter aerogenes, Escherichia coli, Salmonella typhi, Salmonella enterides, Mycobacterium tuberculosi and Shigella dysentariae*. It also inhibits the growth of many fungi. Its action is produced by inhibiting the glutamic acid metabolism of cells. Spirochin, a sulphurated amino base from the root, is a strong prophylactic and antiseptic against a wound infection with gram-positive bacteria, particularly *Streptococcus* and *Staphylococcus*, even when the bacteria are present as a thriving infection. It is experimentally effective in a dilution of 1:70,000. It also promotes growth of the epithelia while providing analgesia and reducing fever. Spirochin also has pharmacological effects on contractions of the heart, intestines and uterus. Athomine is an antibiotic compound from the root bark which shows particular promise in treating cholera. Its potency is intermediate between chloramphenicol and streptomycin. The stem bark contains two alkaloids similar in action to ephedrine which have been used to treat asthma and coughs and as a heart stimulant. A derivative of the mustard oil (benzylisothiocyanate) in the seeds is a very active bactericide against *Bacillus subtilis* and *Mycobacteria phlei*.

A coarse fiber from the bark is used to make mats. A polyuronide called moringa gum, composed of arabinose, galactose, glycuronic acid and traces of rhamnose, is exuded from the injured trunk. It is highly viscous and is used in medicinal formulations and calico printing. A fine lubricant called Ben Oil, with no smell, taste or color, suitable for watches and other delicate mechanisms, is refined from the seeds of this and similar species. The oil has been used medicinally, for soap making and for cooking since classical Greek times. As a lamp oil, it has no scent and burns cleanly with no smoke. It is highly esteemed as a base for perfumes due to its great power to absorb and retain even the most delicate scents. The powdered seeds have strong coagulating properties and are used by indigenous people to induce settling of suspended sediment in turbid and muddy water. The roots contain glucosinolates which are broken down by myrosinase to release mustard oils in a manner similar to that of radish, mustard and horseradish. The roots can be peeled and ground into a pungent sauce similar to horseradish. The mature leaves and twigs are used as livestock fodder. Mature green pods are used as a vegetable in Indian curries, and the young seeds are cooked like peas. Ripe seeds are eaten like peanuts after roasting. Numerous cultivars have been developed for fruit qualities. Jaffna has fruits to 90 cm with a soft flesh and good taste. Chavakacherri has pods to 120 cm (4 ft) long and Palmurungai has a thick pulp. The young pods, shoots and leaflets are cooked and used as table vegetables in the Philippines.

Folk medicine has used the crushed seeds to treat warts and the seed tea to treat fevers. The ground seeds are used to treat warts and the seed oil to treat skin diseases. The leaves have been used to treat scurvy, dog bite, hysteria and flatulence. The alcoholic extract of the roots is used as a diuretic, to treat edema caused by malaria and kidney malfunction and as a liniment to treat rheumatism. The root tea is used to induce abortions and to treat asthma, inflammations, gout, hoarseness and the bites of rabid animals. The gum which exudes from the injured bark is used to treat boils, earache, headache, flatulence and toothache, and as a pessary to induce abortion. The flowers are considered to be aphrodisiac, and effective in treating asthma, gallstones, kidney stones and worms. The continuous blooming provides bees with nectar to produce a quality honey year-round.

Notes The original Malabar name is used for the genus name. The species is from the Latin word meaning "oil bearing," a reference to the oil derived from the seeds. The species names *moringa* and *pterygosperma* have also been applied (perhaps erroneously) to this tree. The common English name for the tree is derived from the spicy condiment similar to horseradish which can be made from minced roots. Other common names include ben tree and drumstick tree. The leaves are high in protein, with leucine 9.3%, valine 7.1%, phenylalanine 6.4%, isoleucine 6.3%, arginine 6.0%. The leaves are also high in vitamins and minerals. The seed oil is composed of 65% oleic acid, with lesser amounts of palmitic, stearic and behinic acids.

Cajeput
Melaleuca quinquenervia

p. 63
Myrtaceae

Toxic properties Flowers and new foliage give off a volatile compound which produces allergic reactions in some individuals. The skin-irritant cajeput oil in the leaves and stems contains 50% to 60% cineol (also called eucalyptol), along with pinene and terpineol. In lesser amounts, the oil contains limonene, sesquiterpene alcohols and benzaldehyde. This oil has an allergic cross-reaction with eucalyptus oil, which contains up to 80% cineol.

Symptoms The flowers cause respiratory distress, reddening and swelling of the face, a burning rash and headache. The sap may produce inflammation and blistering. Overdose of cajeput oil distilled from foliage causes gastrointestinal upset, kidney inflammation and sometimes nervous system disturbance. Externally, the oil may cause dermatitis and blistering in certain individuals.

Treatment Those suffering from respiratory distress should attempt to stay upwind of flowering trees. Those with a skin reaction to the oil should wash carefully after contact with the plant.

Beneficial uses The bark has been used by indigenous people as a fragile but easily constructed canoe. It has also been used for sacred writings, and as an impromptu mattress, a resilient packing material, floats for nets and life belts, furniture stuffing and thermal insulation. The shredded bark is a useful component of potting soil when mixed with peat. The attractive reddish-brown wood has been used ornamentally as flooring, cabinets and gun-stocks and in carving. It is termite-resistant and durable when used in the ground as posts, railroad ties or foundations and in the water as dock members or ship timbers. The cajeput oil distilled from the leaves and twigs is listed in the *British Pharmaceutical Codex* as a liniment for rheumatic joints and internally as a stimulant. Folk medicine uses it internally to treat hysteria, fevers, flatulence and rheumatism. Tiger Balm, an ointment widely used in the Orient to treat arthritis and rheumatism, is reputed to be a mixture of cajeput oil, oil of wintergreen and turpentine. It has been used medicinally to treat intestinal worms, toothache and fungal and parasitic skin diseases such as scabies. The oil is also used as a mosquito repellent in a manner similar to citronella and to eliminate external parasites of man and animals.

Folk medicine has used the leaf tea to clear the lungs, as a stimulant and to help induce sleep. The oil has been used in folk medicine to treat acne, bronchitis, burns, colds, diarrhea, earache, eczema, gout, laryngitis, malaria, rheumatism, skin diseases, scurvy, sprains and tumors. The abundant amber honey produced from the tree has a pungent taste considered to be unpleasant by many. More than 5% presence of *Melaleuca* may ruin other desirable honeys. Because the flavor component is volatile, most of it can be eliminated by allowing the honey to sit in the sun, by vigorous stirring or by gentle heating. There is some demand for the honey from those who believe that eating it will confer an immunity to the respiratory irritation caused by the blooming trees.

Notes The generic name is derived from the Greek words for "black" and "white," referring to the black stem and white branchlets of one of the other species. The species name refers to the usual 5 veins of the leaves. This tree has also been known as *Melaleuca leucodendron*. Other common names for this tree include punk tree, tea tree, bottle-brush and paperbark. Cajeput oil contains cineole, dl-pinene, terpineol and other compounds. The bark contains melaleucin (betulinic acid).

Four o'clock   p. 64
Mirabilis jalapa   Nyctaginaceae

Toxic properties There is some controversy about whether the strong purgative action of the root is caused by the oxymethylanthraquinone it contains, the alkaloid trigonelline or another resinous constituent.

Symptoms Externally, the sap irritates the skin and mucous membranes. Consumption of the flowers, fruit sap or root products induces vomiting and diarrhea. Handling the seeds or roots has caused dermatitis.

Treatment The sap should be immediately washed from the skin or rinsed from the eyes with clean water. The digestive upsets produced by excessive consumption are self-limiting.

Beneficial uses This plant is widely cultivated as an ornamental. Trigonelline extracted from this plant has been found experimentally effective in treating cancer of the cervix and liver. It also reduces blood sugar and cholesterol in experimental animals. Extracts of the seeds have shown antibacterial activity against the common pathogenic bacteria *Staphylococcus aureus, Streptococcus pyrogenes, Salmonella typhi, Vibrio cholerae* and *Shigella flexneri*. One of the most potent antifungal proteins known has been extracted from the seeds. These proteins show a considerable homology to the neurotoxic peptides isolated from spider venom. A ribosome-inactivating antiviral protein has been extracted from the seeds of the plant. The seed has shown contraceptive activity. The flowers are the source of a crimson dye and are used as coloring in cakes and jellies. The nocturnal fragrance of the flowers is reputed to repel and stupefy mosquitoes. Folk medicine has used the leaves as a poultice on abscesses, bruises, boils and itching skin. Internally, the fresh leaf juice has been used to treat uterine discharges and gonorrhea. The root is used as a diuretic, purgative and tonic. The powdered seed has been used as a cosmetic.

Notes The genus name is derived from the Latin word meaning "wonderful." The species has the previous name of the genus, which refers to the town of Jalapa in Mexico. Common names also include marvel of Peru and beauty of the night. The reputedly nontoxic alkaloid trigonelline has been extracted from the plant. The compounds indicaxanthin, mixaxanthin and several derivatives are found in the flowers. The root contains the sterols beta-sitosterol and stigmasterol. The seed has been used as an adulterant in black pepper.

Mexican poppy   p. 65
Argemone mexicana   Papaveraceae

Toxic properties The alkaloids sanguinarine and dihydrosanguinarine found in the seeds and root are the primary toxins. The physiologically active isoquinoline alkaloids berberine, protopine, coptisine, allocryptopine and dihydrochelerythrine are found in all the plant parts. The total alkaloid toxicity has been tested in the mouse and rat. The results extrapolated to man would indicate a lethal alkaloid dose for a 100-kg (220-lb) man would be 0.1 g (0.004 oz). As a result of testing on monkeys, it has been recommended that maximum allowable argemone contamination of oil for human consumption be less than 0.01%.

Livestock have been poisoned by the inclusion of this plant in hay, but the far more common route of intoxication is the seed's being included as a contaminant of other grains. The seed is common as an adulterant in locally produced mustard seed intended for use as a spice in less-developed parts of the world. Mustard seed oil used for cooking has been adulterated with argemone oil accidentally and intentionally to increase profits. Although sanguinarine is reputed to be the primary toxicant in argemone oil, the oil is 4 to 7 times as toxic as the included sanguinarine would indicate it should be. Experimentally, sanguinarine alone fails to induce the symptoms produced by consumption of the oil. Oil obtained by pressing the seeds is highly toxic, but oil obtained by solvent extraction is nontoxic. Sanguinarine is equally abundant in both oils. Thus, the toxicity may be due in part to the other alkaloids or a synergistic action. A toxic amount of the alkaloid or its degradation products may be transmitted in the milk of humans or animals not yet showing toxic symptoms. The yellow sap is slightly corrosive and produces a dermatitis in sensitive individuals. The prickly leaves and fruit capsules produce minor mechanical injury to the tender skin of humans and the mouths of livestock.

Symptoms The seed oil has caused acute human epidemics of dropsy when it has been included as an adulterant in the preparation of edible mustard and sesame seed oils. Mustard seed oil contaminated with *Argemone* oil and used regularly for body massage has produced symptoms of epidemic dropsy. The early symptoms include aches, with noninflammatory swelling of the lower limbs, sometimes with a redness and burning of the skin. As the affliction progresses, fever, enlarged liver, sarcoma-like swellings of the skin, red spots, bleeding gums, nosebleed, excessive menstrual discharge, diarrhea and increased swelling of the legs become evident. The eyes become affected, with engorgement of the retinal veins, retinal hemorrhage and high-pressure glaucoma, which produces blurred vision and may leave permanent vision defects. The primary pathological change associated with the disease is increased capillary dilation and permeability, with an increased tendency to hemorrhage.

Injection of *Argemone* oil produces increased local vascularization. Experimentally, the oil has been found to disturb liver function significantly. Chronic exposure to low-level contamination of food with sanguinarine is suspected as the cause of high levels of esophageal cancer in India. Chronic administration to experimental animals has resulted in degeneration of the heart, liver, kidney and adrenals. In addition to direct contamination of food, the alkaloid may be obtained in the milk, liver or eggs of animals feeding on sanguinarine-producing plants. Acute cases of dropsy have resulted from the consumption of the root as a folk medicine. The seeds, included as an adulterant in grain, have caused considerable mortality and illness in domestic poultry, which show swelling of the wattles, diarrhea and a peculiar gait in which they lift the legs high without bending them. In chickens, the consumption of 56 g (2 oz) of seed a month results in death. The seeds seem to be consumed with impunity, in modest amounts, by quail and other wild birds.

Treatment Discontinuation of the exposure has been followed with symptomatic treatment consisting of diuretics, antipyretics, anti-inflammatory drugs and multivitamins. Diet deficiencies of vitamins C, D, E and rutin may play a part in development of poison symptoms, but are not useful in treating symptoms.

Beneficial uses The seed yields up to 38% of its weight as oil, called kataka oil, which has useful industrial properties and has been burned as an illuminant. The toxic characteristics of the oil deteriorate when exposed to light and almost completely disappear when the oil is heated to 250°F for 15 minutes. The entire plant has been used as an insecticide. Berberine has a wide range of biological activities. It acts as a tonic in small doses. It has been used successfully in treating chronic gallbladder problems and trachoma. This bitter alkaloid slows the heart rate and makes the heart less vigorous and produces dilation of the peripheral blood vessels, leading to a sharp and persistent drop in blood pressure accompanied by a fall in body temperature. It is analgesic when injected, but leaves a lasting pigmentation at the site of injection. Small doses stimulate respiratory motion, but larger doses produce an initial stimulation followed by depression.

It relieves intestinal inflammation and depresses peristalsis. It is antibacterial and antifungal and has effectively cured cutaneous leishmaniasis. In overdose, death is produced by respiratory failure while the heart continues to beat for an extended period. Protopine isolated from the seed sensitizes the heart to fatal ventricular arrhythmias in the presence of adrenaline. It also strongly stimulates the uterine muscle to persistent rhythmic contractions in concentrations of as little as 1:1,000,000. Allocryptopine occurs in two allotropic forms, with the alpha slowing the heart and bringing good results in patients suffering from auricular fibrillation and flutter. The beta form increases the amount of current necessary to provoke fibrillation and causes increased heart rate, in part due to the drug's antivagal action. The L-glutamic acid present in the defatted seed meal has been used to treat certain mental deficiencies in children. Folk medicine has used a tea from the leaves to treat ailments of the spleen, jaundice, fever, cough and asthma. Tea made from the entire plant has been used as a diuretic and to treat hepatitis, cancer, epilepsy and fevers. The plant sap has been used to treat corneal opacities, eczema, skin ulcers, snakebite, ringworm, warts and wounds and as a purgative.

It seems bizarre, but the plant juice has been used in India to treat dropsy, jaundice, skin problems and eye conditions. The seed has been used as an emetic and laxative and to treat diarrhea and dysentery. The seed oil is used as an emetic, purgative and sedative as well as to treat colic, cholera and snakebite. The juice, leaf and seed are reputed to have narcotic and hypnotic properties. Triglycerides from the seeds have been found effective in controlling nematodes of plants. The leaves are smoked as a euphoriant, producing a mildly narcotic analgesia. Bees eagerly visit the flowers of this plant to collect pollen, but do not produce much honey.

Notes The generic name is from the Greek *argemon,* meaning "cataract of the eye," for its reputed medical use in treating this ailment. The species refers to the plant's origin from Mexico. This plant is also known as prickly poppy, thistleroot and donkey thistle. It produces seeds and germinates prolifically, and once established is often persistent and spreading. Modern herbicides give moderate control of seedlings, but vigorous seed production and continuous germination throughout the growing season make it difficult to eliminate. The bacterial wilt *Xanthomonas papavericola* controls the spread of the weed in some areas. Nontoxic dihydroxysanguinarine may make up over 85% of the alkaloid in the seed oil, but the 5% sanguinarine accounts for the majority of the toxic effects. Berbine is probably the same chemical as argemonine, decomposing at 160°F. Protopine, with the same formula but a different structure, melts at 207°F.

Love in the mist p. 66
Passiflora foetida Passifloraceae

Toxic properties The young leaves and green fruits contain a cyanogenic glycoside. The fruits contain the alkaloid 5-hydroxytryptamine and the leaves contain an unnamed alkaloid.

Symptoms Although animals seldom eat quantities of the plant, cyanide poisoning in livestock, with rapid shallow breathing followed by weakness and staggering, has resulted from its consumption.

Treatment In acute cases, intravenous injection of a solution of sodium thiosulfate and sodium nitrate can bring dramatic relief by neutralizing the hydrocyanic acid in the tissues.

Beneficial uses Young leaves are cooked and eaten with rice in Indonesia. Folk medicine uses the leaf tea to treat stomach upset, vomiting, inflammation of the kidney or bladder and female problems. Tea prepared from the vine has been used to treat eczema, eye inflammations, kidney problems, measles, prickly heat, rashes, skin ulcers, urinary burning and wounds. The flavonoid ermanin extracted from this plant has been found in experiments to inhibit growth and deter feeding in several insect pests. The seeds and pulp of the ripe fruit are highly esteemed by many species of birds. The sweet pulp of the ripe fruit is reputed to be edible.

Notes The generic name is derived from Latin words for "passion" (referring to the crucifixion) and "flower." The species name is derived from the Latin word meaning "stinking," referring to the strong and disagreeable smell of injured foliage. One of the common names is stinking passion flower. The alkaloid methyl-beta-carboline, a plant growth inhibitor, has been isolated from this plant.

Pokeweed p. 66
Phytolacca americana Phytolaccaceae

Toxic properties The alkaloid phytolaccin is the primary toxic component in all parts of the plant. Pokeweed root and seeds contain the bitter, acrid triterpene saponins phytolaccagenin and phytolaccatoxin. The latter is similar to the cicutoxin of water hemlock. The root contains an acidic steroid saponin reputed to be primarily responsible for the root toxicity. Humans most often are poisoned after consuming new shoots or the green berries. Fatal poisoning has resulted from eating the ripe berries, but other reports claim that consumption of the fully ripe berries produces no adverse effects and that they have been eaten with impunity in fruit pies. Poke contains a proteinaceous mitogen which produces blood cell abnormalities when absorbed into the body. Livestock are particularly prone to graze pokeweed in intoxicating amounts when it is succulent in the spring or later when pasture is sparse. Pigs are inclined to dig up and eat lethal amounts of the root. The root is the most toxic part of the plant and has caused human fatalities when eaten intentionally or when mistaken for parsnip or horseradish. When the root is powdered for medicinal use, the dust is very irritating to the eye and produces sneezing and coughing.

Symptoms In humans, the symptoms of burning mouth, throat and stomach, vomiting, diarrhea, intestinal cramps, blurred vision, dizziness, headache, sweating and great fatigue with continuous yawning may be present within 30 minutes or take up to 5 hours to develop. Low blood pressure and reduced respiration are present in severe cases and may lead to death. Recovery is usual within 48 hours. In livestock including cattle, horses, sheep and hogs, vomiting and diarrhea may be followed by convulsions within a few hours. If a large quantity of leaves or roots is eaten, respiratory paralysis may result in death. Chronic consumption results in a decrease in milk yield and diarrhea in cattle. Postmortem examination of livestock may show hemorrhage in the gastrointestinal tract and liver of animals with prolonged sickness, but acute poisoning produces only a mild gastritis. Poultry eat the ripe berries and develop an unpleasant flavor, with flesh that produces purgative effects when eaten. Turkey poults develop enlarged, crooked hock joints and an inability to walk and sometimes die from eating the berries. The gallbladder of poisoned birds contains a brownish fluid. Contact with the green plant or the root may cause inflammation of the skin in sensitive individuals. The juice of the plant, a strong root tea or an ointment made from the plant produces a sensation of heat on the skin.

Treatment The stomach contents should be removed and an activated charcoal slurry administered. Demulcents such as milk, egg white or vegetable oil may help soothe and protect the intestinal tract. In humans, treatment for intense abdominal pain may be needed to prevent secondary shock. Fluid and electrolyte replacement may be necessary. There is no specific treatment for pokeweed poisoning in livestock, but tannic acid has been recommended as being helpful.

Beneficial uses Pokeweed mitogen in very low doses has been found to stimulate mitosis (cell division) by a factor of 4 in T (thymus derived) lymphocytes. This effect could have wide therapeutic value in human medicine. Phytolaccagenin is very effective when used to control mollusks and external parasites. The practical application of this molluscidial property was discovered when dead snails of the species which transmit schistosomiasis were found downstream of an area where native Ethiopians were doing their laundry using *Phytolacca* berries as a soap. Inadvertently, the people had been protecting themselves from schistosomiasis by selecting this plant to help clean their clothes.

Phytolaccatoxin has toxic properties similar to those of the cicutoxin of water hemlock. The young shoots and leaves are commonly eaten in the spring as greens called poke salat after thorough cooking and discarding of the first (and sometimes second) cooking water. The young shoot, treated in a similar way, has been used as a substitute for asparagus, with care needed to exclude even the tiniest piece of root. While new growth has significant vitamins and minerals, the social tradition of consumption in the spring would seem imprudent, as poisonings continue to be reported. The fruit juice, containing the betacyan pigment caryophylline, has been used as a dye and to color food and wine. A diluted preparation of the leaf has been found effective in inactivating tobacco mosaic virus. The steroid saponin from the root has been tested and found to be an effective contraceptive. Folk medicine has used the root to induce abortions, treat hemorrhoids and as an emetic and purgative. The berries have been used to treat inflammation, snakebite, syphilis and skin diseases. A wine produced from the ripe fruit has been used medicinally to treat arthritis and rheumatism. The foliage tea has been used to treat conjunctivitis, indigestion, rheumatism and scurvy. Historically, the root was listed in the U.S. Pharmacopoeia as a treatment for rheumatism. Homeopathic medicine still uses the plant to treat several ailments and the root to treat rheumatism. One recipe for rheumatism treatment calls for 1 teaspoon of the tea produced from boiling a tablespoon of chopped root in a pint of water. Fatalities have resulted from overdoses, yet as recently as 1993 a book on herbal medicine suggests the use of the dried root or the tincture as a "lymphatic cleanser." Many wild birds and mammals feed on the ripe berries with seeming impunity.

Notes The genus is derived from the Greek *phyto,* meaning "plant." The second part of the word has been reported as being Latin, derived from the Hindi *lakh,* referring to the dye properties derived from the lac insect. It has also been ascribed to the Greek *lacus,* meaning "lake," describing the blue-purple of the fruits and vegetative parts of the plant. The species name is derived from its American origin. This plant is also called *Phytolaca decandra,* which refers to the 10 anthers in its flowers. Other common names for this plant are inkberry and pokeberry. The toxin is cumulative, as evidenced by a report of diarrhea in humans caused by consuming pigeons which had been feeding on the berries.

Bloodberry

p. 67

Rivina humilis

Phytolaccaceae

Toxic properties The toxin in the roots, leaves and fruits has not been identified, but may be similar to that of *Phytolacca*.

Symptoms Consumption of the fruit produces numbness of the mouth within 2 hours, with a feeling of warmth in the throat and stomach. This is followed by coughing, thirst, tiredness with yawning, and subsequent vomiting and diarrhea (sometimes bloody). The leaves and roots contain larger amounts of toxin.

Treatment The digestive tract disturbances are usually self-limiting. The stomach should be emptied if this has not been accomplished naturally. The intense abdominal pain may require analgesia, and support should be provided to prevent development of secondary shock. In severe cases, fluid replacement may be required.

Beneficial uses This plant is cultivated as an ornamental and at one time was grown to use the fruit juice as a dye and as an ink. Folk medicine has used the plant tea to treat colds, diarrhea, difficult urination, flatulence, gonorrhea, jaundice and ovarian pain.

Notes The genus is named in honor of August Rivinus, a 17th-century German botanist. The species epithet is from the Latin meaning "lowly" or "dwarfish," referring to its small size and sprawling growth habit. This plant is also known as rouge plant, coral berry, baby pepper and pigeon berry. Rivianin, derived from the fruit, is very similar to the red pigment in beets.

Sudan grass

p. 68

Sorghum bicolor

Poaceae

Toxic properties Under certain weather conditions, the young or injured plants have a high content of the cyanogenic glycoside dhurrin, which releases hydrocyanic acid after consumption by livestock. Under certain other conditions, these plants have accumulated toxic levels of nitrates. Young dark-green growth under 60 cm (2 ft) and plants growing on soil with high nitrogen and low phosphorus are likely to have higher levels of toxicity. Second growth after pasturing or removal of hay is particularly dangerous.

Symptoms Livestock seem apprehensive, with rapid, shallow breathing followed by weakness, staggering, convulsions and coma leading to death. Symptoms are often brought on by the consumption of water after grazing on the toxic plants. Progression is often rapid and may lead to prostration within 10 to 20 minutes after the first noticeable symptoms.

Treatment   Prompt treatment by a veterinarian for cyanide poisoning is essential and usually effective. The intravenous injection of a solution of sodium thiosulfate and sodium nitrite can bring dramatic relief. Typical dosage is 660 mg/kg sodium thiosulfate with 22 mg/kg sodium nitrite for large, lethal doses of cyanide. The sodium nitrate releases the cyanide from the tissues, which then combines with the thiosulfate to form thiocyanate, which is readily excreted in the urine.

Beneficial uses   Numerous cultivars are raised as grain, making it the fourth most important world cereal after wheat, rice and maize. The seeds have high nutrient value, with up to 15% protein and 5% fat making them considerably more nutritious than most other grain. This plant is very drought-resistant; throughout Asia and parts of Africa, it is grown when rice and other more preferred crops cannot thrive. The seeds are locally ground and made into bread, tortillas or porridge. They may also be malted and made into various types of beer (which serves as a valuable dietary supplement due to the high B-vitamin content), popped like popcorn or roasted and used as a coffee substitute or extender. The starch from the seeds is used in sizing textiles and paper, as an adhesive for stamps and envelopes, and as a thickening agent for pies and gravies in industrial baking. One subspecies with a sturdy panicle and long, straight side branches is used to make brooms. Several others are cultivated for the sugar content of their stem's sap, which is extracted and boiled to make a syrup. The press cake remaining after sugar extraction contains 10% wax. The grain is used in many areas as a livestock feed. The stem is used for thatching, baskets and fish traps and as a source for a red leather dye. Furfaural, an artificial flavor used in commercial food preparation, is derived from the stems or threshed ears. Folk medicine has used the seeds as a food, tea or poultice to treat various chest ailments. The root is used as a malaria remedy.

Notes   The genus is derived from the Italian name *sorgo*, used for this plant, but the origin is obscure. The species epithet refers to the two colors on the seeds. This plant has also been called *Andropogon sorghum*, *Sorghum sudanensis*, *Sorghum vulgare*, *Sorghum dura*, *Holcus sorghum* and *Sorghum saccharatum*. Some other common names are Guinea corn, kaffir corn, sorghum, millet and milo. Per 100 g serving, sorghum grain has 287 mg iron, 0.38 mg thiamine, 0.15 mg riboflavin and 3.9 mg niacin. The syrup prepared from sorghum has more iron, riboflavin and niacin than cane or maple syrup.

Johnson grass

Sorghum halepense

p. 68

Poaceae

Toxic properties Members of this genus contain the cyanogenic glycoside dhurrin. When plant cells are damaged by wilting, frost or chewing, various enzymes are released which break down dhurrin and release hydrocyanic acid (prussic acid). The toxin prevents the cellular uptake of oxygen from the blood. A cow may be killed by consumption of 2 kg (5 lb) of wilted grass and a sheep by 20% of that amount. Thus, a relatively innocuous grass can be turned into a rapidly lethal poison when the stress induced by frost, drought or mechanical damage induces wilting. Young growth and plants growing on soil with high nitrogen and low phosphorus are likely to have higher levels of toxicity.

Symptoms Livestock seem apprehensive, with rapid, shallow breathing, dilated pupils and muscle tremor followed by weakness, staggering, gasping, convulsions and coma leading to death. Symptoms are often brought on by consumption of water after grazing on the toxic plants. Progression is often rapid and may lead to prostration within 10 to 20 minutes after the first noticeable symptoms. Postmortem examination shows a highly oxygenated, bright-cherry-red blood which clots very slowly.

Treatment Prompt treatment by a veterinarian for cyanide poisoning is essential and usually effective. The intravenous injection of a solution of sodium thiosulfate and sodium nitrite can bring dramatic relief. Typical dosage is 660 mg/kg sodium thiosulfate combined with 22 mg/kg sodium nitrite for large, lethal doses. The sodium nitrate releases the cyanide from the tissues, which then combines with the thiosulfate to form thiocyanate, which is readily excreted in the urine.

Beneficial uses This grass can be used as a safe source of animal feed as hay or silage. In the case of silage, it should remain in the silo a minimum of 6 weeks before use. With careful management, it has been used as a forage crop. The grain may be ground into a flour for use as human food.

Notes The genus is derived from the Italian name *sorgo,* used for this plant. The species name is derived from Allepo, its place of first collection in Syria. This plant has also been known as *Holcus halepensis* and *Andropogon halepensis.* The common name refers to Colonel Johnson, a farmer in South Carolina who first grew and distributed this grass in 1830 after a friend brought him some from Turkey. Another common name for this plant is Egyptian millet.

Akee   p. 69

Blighia sapida   Sapindaceae

Toxic properties The seeds contain the cyclopropanoid amino acid hypoglycin A and its dipeptide derivative hypoglycin B. The unripe arils contain hypoglycin A and are the primary source of human poisoning. As the names of the compounds indicate, they induce a catastrophic fall in blood sugar, accompanied by glycogen depletion in the liver and other complex metabolic disturbances.

Symptoms Consumption of the unripe fruit (before it splits open and yawns) or the accidental inclusion of a small, undeveloped seed in the food produces "vomiting sickness." The disease is characterized by an initial bout of vomiting 6 to 48 hours after eating the fruit, followed by quiescence and often sleep for several hours. Recurring vomiting bouts accompanied by severe hypoglycemia and depletion of liver glycogen lead to exhaustion, dehydration, coma and death. The average duration of symptoms before death is 12 hours, although some patients may survive as long as 4 days before expiring. Mortality rate is reported as being 40% to 80%, but it is probable that not all cases are reported. For those who survive, convalescence is usually complete within 24 hours of the first onset of symptoms. Although a transient rise in temperature and passing of loose stools may occur, diarrhea and fever are generally absent. Depressed body temperatures are common in advanced cases of toxicity. The lethal effects of hypoglycin A are enhanced in rats which are undernourished or on a low-protein diet. Human case reports seem to parallel this phenomenon. A riboflavin deficiency is shown in the majority of patients with vomiting sickness. Patients are often underweight and have signs of a protein deficiency. Livestock poisoned from eating ackee fruit show a very high mortality rate.

Treatment Stomach contents should be removed immediately if vomiting has not done this. Intravenous glucose can restore normal blood glucose levels, but does not always prevent death due to related metabolic disturbances. Riboflavin partly protects against chronic poisoning and may facilitate replacement of inhibited enzymes or serve as an antagonist of hypoglycin. Through complex biochemical pathways, glycine at the rate of 250 mg/kg/day intravenously limits the toxic effects of hypoglycin in rats. There is no specific published antidote for the toxin in humans.

Beneficial uses The ripe arils of the fruit after it has split open contain no detectable hypoglycin and are highly favored as food, particularly in Jamaica. They are eaten raw, boiled, fried, roasted or used in soups and stews. The classic Jamaican preparation is salt-cod and ackee. The arils are canned in brine and exported from Jamaica. The fruits develop a lather in water and are used for washing clothes and poisoning fish. Folk medicine has used the leaf and bark tea to treat colds, fevers, upset stomach and gum infections. The arils in various forms have been used to treat epilepsy, migraine, smallpox, tumors, yaws and yellow fever. With showy fruits and attractive foliage, it makes a handsome ornamental suitable for planting in parks and along streets. The light reddish-brown, hard wood has been marketed under the name achin and has been used for beds, boxes, chairs, oars, house construction and pilings for docks. The wood is not eaten by termites. The seed oil has been recommended for commercial use and contains 5 times the beta carotene content of corn oil. It has been used in soap manufacture. The pulverized fruits have been used to intoxicate fish for capture.

Notes The generic name honors Captain William Bligh, who introduced the tree to Jamaica from Africa in 1776. This is the same captain who brought breadfruit to the West Indies from Polynesia on the ship *Providence* after his first attempt was thwarted by the mutiny on the *Bounty*. The species epithet is Latin, meaning "savory" or "good-tasting." The common name is one used by Africans in the native range of the tree and is also spelled ackee. The aril contains 5% protein, 20% fat and 5% carbohydrate, with significant amounts of beta carotene, thiamine, riboflavin, niacin and vitamin C. When the arils are fried, they have the appearance of cooked brains and are thus sometimes called vegetable brains, or *seso vegetal* in Spanish. The recurrence of accidental intoxication has caused this tree to be prohibited in Trinidad.

Soapberry tree   p. 70
Sapindus saponaria   Sapindaceae

Toxic properties The fresh fruit contains 37% saponin, which accounts primarily for its oral toxicity. The irritant sapintoxin has been isolated from the unripe fruits of a similar species.

Symptoms Contact with the fruits or sap sometimes induces a rash or blisters. The rare cases of human ingestion produce disturbances of the digestive tract.

Treatment The toxic reactions to the plant are usually self-limiting.

Beneficial uses The fleshy part of the fruit has been used like soap. The fruit also yields a bright-yellow dye for wool. A medicinal oil has been extracted from the seeds. The seeds have been used as beads, buttons and marbles. The bark, roots, leaves and seeds have been used to stupefy fish for capture. The dried and pulverized seeds have been used as an insecticide and to kill lice and the scabies mite. The roots, bark and leaves have been used in folk medicine to treat anemia, arthritis, fevers, gout, rheumatism, snakebite, stings of rays, tumors and other ailments. The fruit has been used to treat asthma, jaundice, pellagra and psoriasis, and as an emetic and purgative. The light-brown heartwood is hard, heavy and coarse-textured and has been used in carpentry, but is not durable when exposed to the weather. Bees vigorously collect the nectar for honey in the winter.

Notes The generic name is the Latin word *sapindus,* meaning "soap." The species epithet is Latin for a plant known as soapwort. This plant has also been known as *Sapindus marginatus* and *Sapindus indicus.* The many other common names in Spanish, French and Dutch are related to the soapiness of the fruits.

Day-blooming jasmine
Cestrum diurnum

p. 70
Solanaceae

Toxic properties The ripe berries contain a compound similar in action to atropine. The unripe berries contain solanine. The leaves contain the alkaloids nicotine and nornicotine in small amounts and saponins with an action similar to the cardiac glycoside ouabain. The leaves contain a substance which induces calcification of elastic tissues and elevated plasma calcium levels. The overall symptoms are similar to those of vitamin D_3 poisoning.

Symptoms In humans, rapid heartbeat and hallucinations result from fruit consumption. In animals, elevated temperature, labored respiration, rapid heartbeat, salivation and lack of muscle control are common. An acute dose of the leaves produces heart stimulation, followed by decline and failure due to ventricular arrest. Chronic consumption of the leaves by livestock produces a weight loss over several months, with calcium deposits in the tendons, ligaments and major arteries. There is usually a reluctance to move due to the lameness produced by calcium deposits. Even the bones have excess calcium deposited in their interior structure. Consumption of as little as 3% cestrum leaves in the diet produces calcium deposits in the kidneys of chickens.

Treatment Fluid replacement may be required in extreme cases.

Beneficial uses This shrub is often cultivated as an ornamental and for its fragrant flowers. A volatile oil extracted from the leaves has been found to have potent fungicidal properties for human and plant pathogens. The fruit is consumed enthusiastically and with seeming impunity by a wide range of birds.

Notes The generic name is from the Greek word for some other plant. The species epithet refers to the daytime opening of the flowers. Other common names include day cestrum, wild jasmine, woman of the day and Chinese inkberry. The "Manchester wasting disease" of cattle in Jamaica may be caused by consumption of the leaves of this plant. The leaves have been found to contain tigogenin and tigonin.

Night-blooming jasmine
Cestrum nocturnum

p. 71
Solanaceae

Toxic properties The leaves contain the steroidal sapogenins nocturnoside, smilagenin, trigogenin, and yuccagenin and are suspected of containing chlorogenic acid along with the alkaloids nicotine nornicotine, cotine, myosmine and solanine. The unripe berries contain solanine. A saponin extracted from the leaves has an action on the heart similar to that of ouabain with an increase in rate and amplitude of heartbeat which may be followed by ventricular failure. Allergic responses to the flowers are probably associated with chlorogenic acid or a compound associated with it. The fruit produces symptoms similar to atropine.

Symptoms Consumption of the fruit irritates the intestinal tract. In humans, poisoning may include nervous and muscular excitement, rapid heartbeat, hallucinations, difficulty in breathing and paralysis. The powerful fragrance released from the flowers at night causes headache, dizziness, nausea, sneezing and labored breathing in some individuals. The leaves are toxic to livestock, producing elevated pulse rate and body temperature, decreased respiratory rate, dullness and urinary retention. Slight convulsions may lead to coma and death. The consumption of 7 kg (15 lbs) of foliage has killed a cow.

Treatment Fluid replacement may be required in extreme cases of intoxication.

Beneficial uses This shrub is often cultivated as an ornamental and for its fragrant flowers. The crushed fruits are used in various formulations to kill rats and cockroaches. Historically, the fruit extract was prepared as tablets and used as a sedative in nervous diseases such as epilepsy, chorea and hysteria. The seeds have been experimentally effective as insecticides. A water extract of the leaves has been found to be effective in controlling mosquito larvae in laboratory tests and field experiments.

Notes The generic name is from the Greek word for some other plant. The species is Latin referring to the night-blooming habit of this plant. It is also known as lady of the night.

Angel's trumpet

Datura candida

p. 72

Solanaceae

Toxic properties The primary physiologically active component of the leaves, seeds, and flowers is scopolamine, which makes up 30% to 60% of the total alkaloids, with lesser amounts of atropine, noratropine, norscopolomine and meteloidine. One analysis has shown 0.65 mg of scopolamine and 0.2 mg of atropine per flower. The active alkaloid in the roots and stems is primarily atropine. Scopolamine is slowly eliminated from the body, making recovery a prolonged event. The fragrance or some other unidentified volatile component given off by a heavily blooming plant may produce headache, nausea, weakness and dizziness. The vapors have a reputation of producing a deep sleep and are used by burglars in India and Southeast Asia to insure the insensibility of their victims. Caged birds hanging near a flowering bush may be rendered unconscious.

Symptoms Consumption of the flowers produces dilation of the pupils, dry mouth, rapid heartbeat, flushing of the skin and delirium with hallucinations. The symptoms may vary considerably due to the amount of active drug ingested and idiosyncratic individual response. Less frequently, the systolic blood pressure is elevated and the diastolic pressure is depressed. The temperature may be elevated, along with decreased bowel sounds and urinary retention. Onset of symptoms may be within 5 to 10 minutes from consumption of tea brewed from the flowers, within an hour of eating seeds and 3 hours after eating leaves. Quiet hallucinations often alternate with violent excitement. The hallucinations, involving vivid colors, strange shapes and other visual misperceptions, may continue for as long as 4 days. Doses of 1 to 3 flowers are enough to produce hallucinations, but more than 6 flowers is likely to produce the more serious effects of extreme lassitude alternating with seizures. With large doses, flaccid paralysis interrupted by convulsions may lead to death. Some people enjoy the abundant nocturnal fragrance of the flowers, but others experience violent headache, nausea, dizziness and weakness. Sleeping in proximity to a blooming tree is reported to produce an extremely deep sleep, sometimes followed by a lasting semistupor.

Treatment The stomach contents should be immediately removed. With severe intoxications, a slow intravenous drip of physostigmine (2 mg for adults or 0.5 mg for children) or pilocarpine helps relieve the symptoms. The duration of influence of the alkaloids is considerably longer than the duration of action of physostigmine; thus, repeated administration often is required to prevent relapse. Sedation and temperature reduction may be required, depending on the symptoms. Phenothiazines should not be given because they potentiate the effects of the alkaloids and may precipitate cardiovascular collapse and death.

Beneficial uses Scopolamine is used to prevent motion sickness when it is absorbed through the skin from an adhesive patch applied behind the ear. Atropine has been used to counteract the depression caused by morphine. It serves as an antidote for phosphate insecticides such as tetraethylpyrophosphate and the dialkylfluorophosphate nerve gases of chemical warfare. Folk medicine has used the flowers placed beneath a pillow to induce a sound sleep. The cooked leaves have been used as a poultice on tumors, and the fresh leaves have been applied externally to relieve headache. The leaves and flowers have been smoked to relieve asthma and influenza. Various formulations have been used internally to treat arthritis, asthma, fractures, rheumatism, sleep disorders and tumors.

Notes The genus is derived from the Hindi vernacular name *dhattura*. The species epithet refers to the flowers being pure, glossy white. The plant has also been called *Brugmansia candida* and *Brugmansia arborea*. The use of *Datura* as a narcotic plant is recorded in early Greek, Sanskrit and Chinese writing. Honey composed primarily of nectar from *Datura* is intoxicating.

Devil's trumpet   p. 73
Datura metel   Solanaceae

Toxic properties The leaves may contain as much as 0.5% of total alkaloids. Scopolamine makes up about 75% of the toxic alkaloids, which are most concentrated in the leaves prior to flowering of the plant. Atropine, meteloidine, norhyoscyamine and norscopolamine also are present in significant amounts. As the plant matures, the alkaloids shift to the flowers and subsequently become most concentrated in the very young fruit. Deflowered plants have much higher alkaloid content in the leaves. The total alkaloid content and ratio of alkaloids present varies considerably with the growing conditions, with rainy weather lowering the alkaloid content. The related alkaloid cuscohygrine is found in the roots.

Symptoms Ingestion of the seeds, flowers or a tea prepared from the leaves causes a burning, dry mouth, dilated pupils, elevated temperature, hallucinations and incoordination. Drowsiness often alternates with a violent attitude. Moderate intoxication may decline in less than a day, but often leaves the victim in an impaired state, with a loss of memory and confused state for several days. Considerable dilation of the pupils can result from accidentally rubbing the eyes after handling *Datura* plants.

Treatment With severe intoxications, a slow intravenous drip of physostigmine (2 mg for adults or 0.5 mg for children) helps relieve the symptoms. The duration of influence of belladonna alkaloids is considerably longer than the duration of action of physostigmine; thus, repeated administration often is required to prevent relapse. About half of the absorbed dose of atropine is excreted unchanged in the urine, the balance being in the form of various metabolites. Over 90% of the drug and its metabolites is excreted within 48 hours of ingestion.

Beneficial uses Scopolamine and atropine have been used as premedications in anesthesia and as analgesics. These two compounds also have been used effectively as antidotes for poisoning by organophosphorus pesticides. This plant is listed as an official drug in the Indian pharmacopoeia. Extracts from *Datura* flowers have been used recently as a general anesthesia for major surgery in China. Folk medicine has used the leaf tea to treat fevers and the fresh leaves tied to the head to alleviate headache and asthma. The leaves, root bark and flowers are smoked to relieve asthma to the extent that several commercially produced "asthma cigarettes" enjoyed brief popularity before their other adverse effects took them off the market. The factors inducing asthma have been experimentally reduced by the inhalation of atropine. Traditional Chinese medicine has used the plant as an anesthesia and analgesia, and to treat asthma, boils, rectal prolapse and fungus infections of the skin. Folk medicine has used various formulations to treat boils, convulsions, dandruff, earache, epilepsy, rabies, rheumatism, ringworm, snakebite, syphilis and tumors. A poultice is applied to the breasts to reduce excessive milk secretion. The root is smoked to treat tuberculosis. Extracts of the plant have been used as an insecticide.

Notes The genus is derived from the Hindi vernacular *dhattura*. The species is from the ancient Greek name for a similar plant. This plant is also known as *Datura fastuosa* and *Datura alba*. The common name in several languages is a variant of stinkweed. Other common names include goat-provoking flower, drunken-fairy peach and trumpet flower. The leaf is used as a narcotic and inebriant recreationally and with criminal intent. The seeds have been used regularly in criminal poisoning in Asia. The plant has been used to lure virgins into prostitution. The women then used the plant as a sedative to knock out and take advantage of their clients. The roots of this and several other cuscohygrine-containing plants are used as sedatives and narcotics in folk medicine. The honey produced from this plant reportedly produces intoxication due to the above-named alkaloids. This and other *Datura* species have been used by primitive people as the agent of a trial by ordeal, in which the accused consume a concoction of the plant and the survivors are considered to be innocent.

Jimsonweed
Datura stramonium

p. 73
Solanaceae

Toxic properties   All parts of the plant contain the alkaloids atropine, hyoscyamine and scopolamine, with lesser amounts of meteloidine and nor-hyoscyamine. The ratio of scopolamine to atropine in the young plant is 3:1, which is reversed after the plant has flowered. The scopolamine content continues to decline in favor of atropine with further growth. The seed contains about 2.7 mg atropine and 0.66 mg scopolamine per gram. Human poisoning is usually due to intentional ingestion of the seed or flower for the inebriant effects or accidental ingestion as a component of cereal products. Meal contaminated with *Datura* seed has caused mass poisoning of soldiers or livestock. Bread baked with jimson seed-contaminated flour contains an undiminished alkaloid content. Overdose due to consumption of the tea from leaves, flowers or seeds is less common. Accidental deaths due to impaired judgment have been reported as a result of *Stramonium* use.

Because of the strong odor and taste of the foliage when fresh, most livestock poisoning is due to the ingestion of this plant mixed with hay or silage, or the seeds mixed with corn or other grain. Pigs are the most sensitive, with 0.27 g (0.009 oz) of seeds per day producing toxic effects. The toxic dose for cattle is 15 g (0.5 oz) per day. Experimental rats showed liver enlargements when seeds made up 0.5% of the diet. The compound gamma-L-glutamyl-L-aspartate, extracted from this plant, has been found to suppress long-term memory of an event if administered 3 to 24 hours after the event. The leaves and flowers sometimes cause a skin irritation.

Symptoms   In humans, the type and level of intoxication depends in part on the method of consumption, with the whole leaves and seeds being a more gradual onset than the tea, in which the active alkaloid is in solution. The old but still-valid depiction of the symptoms is "hot as a hare, blind as a bat, red as a beet, dry as a bone, and mad as a wet hen." The physical symptoms include dilation of the pupils, blurred or dimmed vision, dryness of the mouth and throat, flushing of the skin, rapid heartbeat, headache and urine retention.

Psychological or behavioral symptoms are described in historic accounts as dreadful perturbations of the mind. More modern descriptions include confusion, combativeness, disorientation, absent-mindedness or a ceaseless, maniacal, purposeless activity followed by lethargy. Sexual inclinations are said to be excited, particularly in women, who may show nymphomania. Visual or auditory hallucinations, with the illusion of flying or misperceived colors, are not uncommon. Victims have reported being followed by knee-high red and black spiders or have repeatedly dived into a pond in search of red-eyed dolphins. Coma and death have been recorded from eating the seeds or from drinking a tea made from the leaves. The lack of response of the pupils to light may last as long as a week. Complete amnesia of the event and the arrest of milk secretion are other occasional ancillary symptoms. Chronic low-level consumption of contaminated cattle feed produces a transient loss of appetite, bloat and dry muzzle, which passes in 2 or 3 days. Atropine impairs the eructation reflex; thus, ruminants may be prone to bloat. In more severe cases, rapid pulse and breathing, dilated pupils, restlessness, nervousness and frequent urination are seen. In fatal cases, animals show a weak and rapid pulse, slow irregular breathing and falling temperature, with coma or convulsions preceding death.

Treatment The caffeine in strong coffee may be used as a stimulant until medical aid is available. In humans, vomiting should be induced by the administration of ipecac, followed by activated charcoal and a cathartic such as sodium sulfate. With severe human intoxications, a slow intravenous drip of physostigmine (2 mg for adults or 0.5 mg for children) usually dramatically relieves the central nervous system symptoms. The duration of influence of belladonna alkaloids is considerably longer than the duration of action of physostigmine; thus, repeated administration is often required to prevent relapse. Physostigmine, pilocarpine and arecoline have been recommended as veterinary antidotes. Livestock poisoning can be prevented by removing the plants from pastures prior to grazing and from hay fields prior to mowing. Prevention of seed production soon results in elimination of the plant from the area.

Beneficial uses The dried leaves and seeds were officially listed at one time as drugs in the *U.S. Pharmacopoeia*. The alkaloids are useful in blocking parasympathetic nerves and interrupting bronchial and intestinal spasms. The Eighth Merck Index shows it as being useful in treating parkinsonism and asthma. Atropine has also been used to treat delirium tremens and as an aid in suppressing symptoms of withdrawal in morphine addiction. The Russians have used it to treat angina, asthma, bronchitis and epilepsy. Chinese medicine uses an extract as an anesthetic for major surgery. Homeopathic medicine finds many uses for the plant, some of which are the treatment of apoplexy, burns, delirium tremens, epilepsy, headache, hiccups, meningitis, nymphomania, rabies, stammering, sunstroke, tetanus and typhus. The alkaloids have been used in purified form by modern medicine to treat psychiatric disorders and to prevent motion sickness. The leaves have been widely used in smoking materials to relieve asthma symptoms. Folk medicine has applied the leaf poultice to abscesses, boils, burns, cancers, gout, hemorrhoids, ingrown toenails, tumors, ulcers and prolapsed uterus. The leaf and fruit juice have been used to treat baldness, dandruff and earache. The leaf is used to dye cloth green. A tea made from the leaves and fruits has been used to expel screw worms and other maggots from wounds of humans and livestock.

Notes The genus is derived from the Hindi vernacular *dhattura*. The species name is an old name for this genus, which is derived from the Latin *struma*, meaning "a swelling." This species seems to be identical to *Datura tatula* except for the purplish stem and flower of the latter. The common name is a corruption of James Town, referring to an event in colonial Virginia in which British soldiers consumed the fresh sprouts of this plant as greens, then showed foolish intoxication for several days. Other common names include thorn-apple, devil's apple, mad apple and witch's weed. The roots contain the alkaloids hyoscyamine, cuschohygrine and 7-hydroxy-3:6-ditiglyloxy-tropane. The plant also contains scolopetin. The powdered seeds have been placed in food and the leaves mixed with smoking tobacco to intoxicate intended victims of robbery and rape. The seeds are reputed to increase the sexual desire, particularly in women, sometimes leading to nymphomania. The seeds have been used in occult ceremonies to induce hallucinations. Honey produced from the flowers is reported to be toxic.

Christmas berry
Lycium carolinianum

p. 74
Solanaceae

Toxic properties The leaves have one or more unidentified toxic alkaloids.

Symptoms Livestock have died after eating the foliage. Humans have been subject to repeated vomiting after eating the fruit. No human mortalities have been recorded.

Treatment The attractive, tasty fruit should be avoided. If consumed by children, it should be promptly removed from the stomach.

Beneficial uses This plant is grown as a hardy, attractive shrub. The fruit and root bark of other *Lycium* are listed in the Chinese pharmacopoeia for the treatment of cough in pulmonary tuberculosis, deficient liver and kidney function, diabetes, fever, pneumonia and the restoration of semen. The Chinese have found that *Lycium* promotes the reattachment of periodontal tissue after planing of diseased dental root surfaces. Endangered whooping cranes feed on the ripe berries in their winter range on the Gulf coast. The ripe berries are reported to be edible, with a sweet-salty taste. Perhaps the poisonings reported are from the green berries. Folk medicine has used *Lycium* to treat cancer, dizziness, headache, high blood pressure, hyperglycemia, hypoglycemia and rheumatism.

Notes The generic name was used by the Greeks for a thorny shrub that grows in Lycia. The species is named after the region in which it was first discovered. This plant is also known as *Lycium americanum*. Other common names include matrimony vine, wolfberry and box thorn.

Tobacco   p. 74
Nicotiana tabacum   Solanaceae

Toxic properties The primary toxic alkaloid is nicotine, with the chemically similar anabasine and nornicotine also present. The leaves and fruits contain the most toxin (2% to 3% nicotine), with lesser amounts in roots, flowers and stems. The ratio and potency of alkaloids varies with different species in the genus. The alkaloid may be absorbed by the lungs from inhaled smoke, through the digestive tract when chewed or through the skin when in solution or when the plants are handled. The alkaloid is extremely addictive, whether smoked in various forms, chewed, snuffed or dipped. Users continue with the habit in spite of public education campaigns aimed at revealing the pernicious effects of the plant. Anabasine, pharmacologically similar to nicotine, is the predominant alkaloid in *Nicotiana glauca*. A dose of 60 mg (0.002 oz) is considered lethal in man.

Symptoms Nicotine is an extremely potent poison with a great rapidity of action. Very large oral doses are fatal within seconds and any lethal dose will produce death in minutes. Doses as small as 3 or 4 drops have produced death in adults. The central nervous system is briefly stimulated by nicotine, then depressed. In acute intoxications, convulsions with elevated heart rate and blood pressure usually precede a curare-type paralysis of the muscles, which leads to death by respiratory failure. Inexperienced users of tobacco in any form inevitably feel the effects of mild nicotine intoxication, which may include salivation, nausea and vomiting, loss of equilibrium and sensory disturbances. Tobacco often stimulates active peristalsis of the gastrointestinal tract, resulting in frequent and vigorous evacuations of the bowel.

It has now been firmly established that the regular use of tobacco in any form greatly increases the probability of various forms of cancer. Tobacco poisoning has occurred as the result of use in folk medicine, as an abortion agent, with introduction to food or drink by accident or with criminal intent, by swallowing saliva when chewing or dipping snuff, by children sucking the condensate from old pipes and by application of the plant poultice or extract to large areas of skin. The plant is poisonous to grazing animals in large doses. Hoofed animals show a staggering gait, salivation and trembling, and often lay down. Many livestock have died after gaining access to and eating tobacco in the field or in drying barns. Cattle, horses, oxen, sheep and ostriches have been poisoned when nicotine solution or extract of tobacco has been applied to control insects, lice or ticks. Exposure to nicotine *in utero* has often resulted in abnormalities of limbs and spine in offspring of experimental animals. Household pets have been killed after consumption of tobacco products. Sweetened plug tobacco has repeatedly proven to be an attractive poison for children and animals. Deaths and serious illnesses have been reported due to the use of leaves of *Nicotiana glauca* as boiled green vegetables. *Nicotiana glauca* leaves have also been the source of death when used intentionally or with criminal intent as a component of the native beer called chicha. Mammals have the capability of becoming inured to most of the effects of nicotine with repeated exposure. Human users of tobacco products show reduced physiological response to nicotine after repeated usage. Wild and domestic herbivores may become acclimated to the consumption of plants containing nicotine.

Treatment If the intoxication is by the oral route, contents of the stomach should be removed and the patient should be given activated charcoal and a cathartic. Vascular support and controlled respiratory assistance should be available and provided as needed. Although no antidote is known for lethal doses of nicotine, early supportive care provides a good prognosis. If modern medical care is not available, the administration of tannin in the form of strong tea will reduce the amount of toxin in the gastrointestinal tract by forming insoluble nicotine tannate. Most fatal cases occur within one hour of the onset of symptoms.

Beneficial uses The nicotine and anabasine in extracts from the leaves, combined and separately, have been utilized to eliminate fleas, ticks, and lice in human and veterinary medicine and as agricultural pest-control sprays. Because of its potential toxicity to the host, it has been replaced in in dustrial nations by various synthetics. Nicotine was the original drug used in Cap-Chur gun darts to induce paralysis and immobilize wild animals for capture. First indroduced in the 1960s, its use has been mostly discontinued, as the frequently associated convulsions were considered inhumane. Animals injected with nicotine show a phenomenon called tachyphylaxis, in which they develop a resistance to subsequent injections within minutes. Folk medicine has used the leaf poultice to treat insect bites, headache, lice, mange, ringworm, scorpion sting, snakebites, ticks and wounds. Tobacco has been used in hundreds of formulations of greater or lesser medical efficacy. Some of the ailments to which it has been applied include asthma, boils, dysentery, earache, malaria, rheumatism, scabies, tetanus and ulcers. Tobacco embedded in bait has been used to immobilize fish for capture. Anabasine from *Nicotiana glauca* is only about _ as toxic as nicotine to mammals, but is very toxic to aphids and has been used effectively as an insecticide. *Nicotiana rustica* has been used in various folk medicine remedies, and has been placed inside stockings to ward off leeches when traveling in damp places. The seed contains 25% to 37% of a nontoxic semidrying oil which is effective in preventing metal corrosion. The oil may also be used for cooking, burned as an illuminant, or sterilized and used as a substitute for olive oil in the preparation of injections. The seedcake remaining after oil extraction is free of nicotine and has potential as a cattle feed with 36% protein.

Notes The genus is named for Jean Nicot de Villemain (1530–1600), who introduced tobacco to the courts of Europe. The species epithet is derived from the original native name for the pipe in which the plant was smoked. Other alkaloids found in tobacco leaves include nicoteine, nicotimine, nicotelline, N-methylanabasin, anatabine, methylanatabine, nicotyrine and nornicotine.

Chalice vine   p. 75
Solandra guttata   Solanaceae

Toxic properties All parts of the plant contain toxic atropine-like alkaloids. The spectrum of alkaloids present is similar to that of *Datura*, with noratropine, hyoscyamine, hyoscine, tigloidine, valtropine and others in trace amounts. Other authors report solanine and solanidine as the primary active compounds. All members of this genus are similarly toxic.

Symptoms Consumption of the flowers or a tea made from them or the leaves produces lack of coordination, dilated pupils, dry throat, fever, hallucinations and irrational behavior. A delayed irritation of the gastrointestinal tract may occur after eating the flowers. Respiratory and circulatory failure is seen in severe cases. Deaths have occurred after chewing fragments of the flowers. Dizziness, headache and nausea have resulted from long exposure to the fragrance. The sap produces temporary blindness if it gets into the eye.

Treatment The neurological disturbances from the alkaloids seem to resolve themselves if the patient is kept in a protected environment until returning to normal. The sap should be rinsed from the eye with clean water and the eye should be protected from bright light until it returns to normal.

Beneficial uses This species and several other members of the genus are cultivated as ornamentals. Several Native American groups have used members of this genus to induce hallucinations in religious ceremonies.

Notes The genus is named in honor of the Swedish naturalist Daniel Solander (1736–1782). The species epithet is Latin for "speckled." There is some confusion in the literature over this and similar species. *Solandra grandiflora* climbs to 12 m (40 ft) and the slender part of the flower is shorter than the calyx. *Solandra maxima* climbs to 18 m (60 ft) and sometimes develops aerial roots.

Black nightshade
p. 76

Solanum americanum
Solanaceae

Toxic properties The primary toxins in the green fruit and leaves are a group of glycoalkaloids characterized by solanine (also known as solanine-t and solatunine) and solasodine. Other toxins present in the plant include chaconine, solasonine, solamargine, solanigrine, gitogenin and traces of saponins. The ripe fruit has been shown to contain acetylcholine, and it has a cholinesterase-inhibiting effect on human plasma. Solanine and solasodine have been reported as inducing developmental abnormalities in the fetus. Solanine from potatoes is cited as the cause of the high incidence of spina bifida in Ireland. In Transkei, rural people have a high incidence of esophageal cancer correlated with the use of *Solanum americanum* as a spice or relish in their diet. The plant may show a separate toxicity to livestock due to the sometimes very high nitrate content of the leaves. The concentration of both toxins in the plant increases with maturity, but varies considerably with the strain of the plant and such environmental conditions as soil type and rainfall. There is considerable variation in the literature on the chemistry and pharmacology of this plant.

Symptoms The literature reports two different toxic syndromes within 30 minutes of eating this plant. The first is from the highly irritating effects of solanine on the gastrointestinal tract. Gastrointestinal symptoms include loss of appetite, fever, salivation, nausea, vomiting, abdominal pains and diarrhea. The second suite of symptoms is neurologic and probably due to the hydrolysis of solanine to solanidine by the loss of the sugar component of the molecule in the digestive tract. The remaining chemical, solanidine, is absorbed from the intestinal tract. The effects on the nervous system of 0.2 g (0.007 oz) may include dry mouth and throat, apathy, dizziness, headache, cramps, drowsiness, labored breathing, sweating and increased heart rate (other reports show decreased heart rate) followed by loss of speech, weakness and trembling, dilated pupils, mental confusion, paralysis, convulsions and unconsciousness. Death is usually the result of respiratory paralysis.

Fatal poisoning of horses, cattle, sheep, goats, pigs, rabbits, chickens, ducks and geese from eating the unripe berries or foliage has been reported. In livestock, the symptoms are labored breathing and constipation followed by a dark diarrhea, dry muzzle and slight fever. The reported toxicity may also be from the high nitrate content of the foliage. Livestock with nitrate toxicity may show muscle tremors, staggering gait, reduced body temperature, rapid pulse, frequent urination, collapse with or without convulsions, coma and death. The LD_{50} of nitrate in the feed is approximately 200 mg/kg. More commonly seen chronic effects are decreased milk yield, abortion and disturbed vitamin A metabolism. Grazing on the plant produces an uncoordinated gait and temporary seemingly impaired vision in horses.

Treatment The stomach contents should be removed and replaced with a slurry of activated charcoal. Various demulcents such as milk or vegetable oil help protect the gastrointestinal tract from irritation. It has been reported that the administration of pilocarpine (a parasympathomimetic drug) usually relieves most of the nervous symptoms. Rapid dehydration may require fluid and electrolyte replacement. Symptomatic and supportive treatment are often required until recovery is advanced. At least 78% of the orally ingested solanine is excreted in the urine or feces within 24 hours. For livestock, an intravenous injection of 20 ml of methylene blue (a solution of 10 gm in 500 ml of water) per 100 lb of body weight helps reduce the adverse effects of nitrates.

Beneficial uses The fully ripe fruits are cooked into jams and preserves or eaten raw. The fruits are considered an excellent source of calcium, phosphorus and iron. The young leaves are cooked and eaten after the cooking water with the soluble alkaloids is discarded; but in some areas the plant is considered a deadly poison. Perhaps several chemically different but externally indistinguishable forms of the plant exist. The leaves have been officially sanctioned drugs in the Mexican, French and Spanish pharmacopoeias. The fruits contain diosgenin and tigogenin, sources material for the production of various commercial steroidal drugs. Solanine has been prescribed as a treatment for neuralgia, migraine and gastralgia and as a sedative for Parkinson's disease. Solasodine is antagonistic to the rapid heartbeat produced by adrenaline. A tea from the whole plant is used in traditional Chinese medicine to treat cancer of the cervix and vaginal infections.

Folk medicine uses leaf poultices to treat abscesses, boils, convulsions, erysipelas, headache, neuralgia, rheumatism, sores and toothache. Extracts of the leaves have been used externally to treat abscesses, eczema, leg ulcers, ringworm, sores and other skin diseases, as drops for inflamed eyes and aching ears, as a douche for moderate menstruation and vaginal infections, and as an enema to relieve fever and typhus. It is eaten to treat asthma, cervical cancer, delirium, meningitis, scarlet fever, stomachache and urinary burning and as a sedative or tranquilizer. The ripe berry has been used as an aphrodisiac and to treat heart and liver problems, asthma, diarrhea, eye diseases, rat bite, rabies, urinary discharges and fevers.

Notes The genus is from an ancient Latin plant name used by Pliny. The species name indicates that it is from America. This plant is also called *Solanum nigrum* and *Solanum villosum* in the literature. Other common names include yerba mora, rooster killer, turkey berry, poison berry, wild pepper and devil's tomato. As an ingredient in a West Indian stew, it is known as branched Kalaloo (also spelled calalu). This plant is an alternate host for several nematodes and many virus diseases of agricultural crops.

Italian jasmine
Solanum seaforthianum

p. 77
Solanaceae

Toxic properties The fruits contain solanine. Careful analysis has shown the alkaloids solasodine, solanidine, solanoforthine and other as-yet-unidentified alkaloids to be present in the stem bark.

Symptoms Consumption of the fruits has made children ill, with loss of appetite, diarrhea and general malaise. The fruit has been reported as intoxicating or lethal to poultry, rabbits, sheep, pigs, cattle and kangaroos.

Treatment The plant remains should be removed from the stomach to prevent further absorption of toxins. In severe cases, supportive therapy in the form of fluid and electrolyte replacement may be needed.

Beneficial uses The plant has been found to control high blood pressure. Solanidine has been shown experimentally to prevent insect feeding.

Notes The genus is from an ancient Latin plant name used by Pliny. A cultivar with white flowers has been propagated. This plant is also known as Brazilian nightshade.

Sky flower
Duranta repens

p. 77
Verbenaceae

Toxic properties The ripe fruits contain a pyridine alkaloid and a saponin. The leaves contain a saponin. A glycoside has been isolated from the bark.

Symptoms Consumption of the fruit produces drowsiness, elevated temperature and pulse rate, swelling of the lips and eyelids, dilated pupils and convulsions. Gastrointestinal irritation and erosion is sometimes present. Circulatory disturbances, neurological symptoms and convulsions provide a grave prognosis. The leaves are bitter and not usually consumed by livestock, but pigs have died after grazing on the plant. Human deaths have resulted from consumption of the fruit.

Treatment The plant material should be removed from the stomach as soon as possible. Fluid and electrolyte loss may need to be corrected and convulsions may be controlled with diazepam. If circulatory disturbances are evident, exchange transfusions may offer the only hope.

Beneficial uses The durable, light-brown, hard wood is used for tool handles and turnery. Birds seem to consume the fruits with impunity and raccoons have been seen sampling the ripe fruit. The plant has been used as a detergent and an insecticide. The juice of the berries in solution is lethal to mosquito larvae. Folk medicine has used the flower tea as a stimulant and the fruit tea to treat malaria and other fevers.

Notes The genus is named to honor Castore Durante (c. 1529–1590), a Papal physician and botanist. The species epithet is from the Latin word meaning "creeping." The plant has also been called *Duranta plumieri*. Other common names include golden dewdrop, lilac, pigeonberry and several Spanish equivalents. The compound lamiid and 3 of its esters with cinnamic acid (called durantosides) have been extracted and isolated from the plant.

Lantana p. 78

Lantana camara Verbenaceae

Toxic properties Consumption of the green berries has caused severe poisonings and death in children. The mode of action of the unidentified toxic compound is similar to that of the belladonna alkaloids and is different from that in the foliage, which affects livestock. The foliage contains lantadene A (also called lanthanine and rehmannic acid) and lantadene B as well as other pentacyclic triterpenes in various reduced and hydrolyzed forms. Most of these compounds inhibit bile secretions and produce other liver malfunctions in livestock. The lantadene is converted by the liver to phylloerythrin, which causes photosensitization, kidney congestion, liver failure and death. Toxicity is not cumulative, but is the result of the consumption of a large amount of the plant in a short period of time. The LD_{50} of lantadene A intravenous in sheep is 1 to 3 mg/kg. Lantadene A is poorly absorbed from the digestive tract, with the LD_{50} orally in sheep being 60 mg/kg. Lantanine B reduces circulation and depresses the temperature. In sheep, the consumption of 10 g (0.35 oz) of the leaves will produce toxic symptoms. The prickly stems may cause an allergic skin irritation in some humans.

Symptoms In humans, the typical symptoms developing 2 to 6 hours after ingestion of the green berries may include loss of coordination, lethargy, photophobia, vomiting, diarrhea, labored slow breathing, cyanosis, dilated pupils, depressed reflexes, coma and (rarely) death. Some individuals find the smell of the leaves to be offensive and develop headache and giddiness. The leaves of this plant are responsible for considerable economic loss due to death and debilitation of livestock. Poisoning by the leaves has been demonstrated in cattle, sheep, water buffalo, goats and rabbits. Some dogs develop a fondness for the plant and will persist in eating the leaves even when admonished not to. Ruminant livestock stop grazing and develop a transient, stinking diarrhea, then become constipated within a few hours.

Recent research has shown that lantana toxins produce stasis of the rumen and the rumen microbes do not metabolize the toxins. The toxins are absorbed directly from the rumen and enter the liver via the portal circulation. Liver damage initiates the hepatoruminal reflex, causing further rumen stasis and absorption of additional toxins, which continue to damage the liver. Within the next day or two, the animal becomes sedated with a fever, and exposed skin becomes photosensitive. In chronic exposure, the skin may be yellow, swollen, hard and cracked, with open raw areas. The photosensitivity is secondary to liver injury. Jaundice is usually evident, due to the inability of the liver to excrete bilirubin. Death is from kidney or liver failure and starvation due to stasis of the bowel. If death does not occur, recovery is very slow. The compound lancamarone from the leaves is toxic to fish at very low doses.

Treatment In humans, supportive medical care, including gastric lavage followed with activated charcoal and saline cathartic, fluid replacement and respiratory support, should be provided as needed. In ruminants, large doses of activated charcoal dispersed in large volumes of electrolyte solution interrupt the cycle of absorption of the toxins and promote rumen detoxification. Sheep should be treated with 500 g of activated charcoal in 4 l of electrolyte solution. Cattle should be treated with 2 kg of activated charcoal in 20 l of electrolyte solution. Treatment with fluids or purging will promote the movement of plant material into the intestinal tract, where further absorption will take place, unless charcoal is present to adsorb the toxin. The charcoal slurry should be given by stomach tube as soon as possible, but is still effective after symptoms have fully developed. Purging the animal without inhibiting the adsorbtion of additional toxins has been shown to be detrimental, with increasing plasma bilirubin levels and frequently, death. The recovering animal should be kept in the shade for several days until the photosensitive symptoms decline.

Beneficial uses This plant has many cultivated varieties which are used as ornamentals in the house and garden. It is not uncommon for the fully ripe fruits to be eaten fresh with no adverse effects. The fruit is also preserved and used as a condiment. Wild birds actively seek the ripe fruit. The bark of the stem and roots contains lantanine, which reduces fever and spasms. The twigs and stems are used as a fuel source and have been recommended as a source for pulp used in paper making. Mixed with newspaper, they are a suitable substrate for mushroom culture. Oil from the seeds is antimicrobial and anthelmintic. Oil from the leaves is antiseptic and toxic to fungi. Folk medicine has used the leaf tea to treat anemia, asthma, chickenpox, colds, diarrhea, fevers, high blood pressure, itch, jaundice, leprosy, measles, mumps, rheumatism, scabies, snakebite, tumors and upset stomach. The plant is used in Chinese medicine to treat malaria, mumps, rheumatism, fevers, scabies, tetanus and traumatic injuries. The root tea has been used to treat asthma, mumps and venereal diseases. The pounded plant has been used as an intoxicant to capture fish. The leaf extract has been used effectively as an insecticide to protect several agricultural crops and to kill mosquito larvae.

Notes The genus is named for the similar appearance to the plant *Viburnum* *lantana.* The species epithet is a South American vernacular name for a species in this genus. An alternative attribution of the origin is the Greek *kamara,* "a vaulted chamber," in reference to the cells of the fruit. Other common names are wild sage, red flowered sage, yellow sage, wild sage, shrub verbena, stink grass and prickly lantana. This introduced lantana was considered to be a troublesome weed over 100 years ago in Bermuda. The similar *Lantana involucrata* forms shrubs with purple flowers and has many of the same properties. The plant also contains cineol, citral, dipentene, eugenol, furfural, geraniol, linalool, phellandrene, phellandrone, terpenol and the triterpenes alpha-amyrin and beta-sitosterol The leaves contain 3-ketoursolic acid, lantanilic acid, lantanolic acid and lantic acid. Lantana exerts an allelopathic influence, restricting the growth of surrounding vegetation. Populations of the weed are reduced by burning the habitat in the dry season. Biological control of this plant has been proposed using a lace bug, *Teleonemia scrupulosa,* which eats the leaves; a fly, *Ophiomyia lantanae,* which destroys the seeds; and a caterpillar, *Hypena strigata,* which defoliates the plant.

Animals

Fire sponge, "Don't touch me" sponge
Tedania ignis, Neofibulia nolitangere

p. 80

Porifera

Toxic element Poisoning is most common after a vigorous physical contact with the sponge which allows the sharp spicules in the sponge to abrade the skin and facilitate the entry of the toxins. The toxin seems able to penetrate undamaged intact skin, as these sponges handled out of the water may produce dermatitis. The specific toxin has not been identified, but sponges are known to produce a variety of unique toxic compounds, including a polybrominated dioxin, four different indoles, a beta carboline, a carbazole, numerous diketopiperazines and 2 substituted benzenes. A *Micrococcus* bacteria isolated from this sponge has been found to produce diketopiperazines in laboratory culture media and may be the source of these compounds in the sponge. A fire sponge in a bucket of water with other marine specimens such as fish, crabs and mollusks will usually kill the other organisms in less than an hour, indicating that the sponge may release a water-soluble toxin. Alternatively, several of the chemical compounds originally ascribed to this sponge have been found to be produced by commensal and symbiotic microflora associated with the sponge.

Symptoms The symptoms of contact with the sponge are an initial burning which may take some minutes to develop. This may be followed by itching, redness and swelling at the site of contact. The discomfort and irritation usually decline after about a week. The irritant chemical has not been definitively confirmed. Some experiments indicate the irritant is simply a toxic chemical introduced into the skin by damage from the spicules, but other studies seem to indicate a contact allergic sensitivity to the sponge substance. The inflamed area may peel after several weeks or the irritation may persist for months after the contact. Systemic anaphylactic symptoms of weakness, sweating, malaise and numbness of the extremity are known but rare. Certain individuals seem to be totally immune to the effects of the sponge.

Treatment Vinegar soaks seem to be the only effective local treatment. Antihistamines, corticosteroids and topical anesthetics are usually ineffective and sometimes aggravate symptoms.

Beneficial Uses Several cytotoxic and tumor-inhibiting compounds have been extracted from this sponge.

Notes The sponge *Neofibulia nolitangere* is also known as *Fibularia nolitangere*. The species name is from the Latin "don't touch me."

Feather hydroid
Gymnangium longicauda

p. 81
Hydrozoa

Toxic element The toxic element is a venom injected by special cells called nematocysts, which fire a venom-filled harpoon upon contact with a food item or a careless human.

Symptoms The potency and volume of the venom vary considerably between species. Some produce a mild sting only on sensitive skin such as the inside of the forearm. Other species produce a more sharp, burning, itching pain when contacted.

Treatment There is no known treatment for the sting, but it is usually self-limiting and of less than a day's duration. Ice is usually an effective analgesic. Heat and sunburn increase the irritation.

Notes The mysterious stings received from brushing contact with pilings and submerged mooring lines are usually from hydroid species. *Lytocarpus*, the Christmas tree-shaped hydroid often found growing on shallow-water structures and inside marine caves, has a sharp sting which often catches novice divers unawares. Even after being stung, many do not recognize the delicate and graceful colony as the source of their discomfort.

Fire coral
Millepora alcicornis

p. 82
Hydrozoa

Toxic element The toxin from the nematocysts has been studied and found to be several protein compounds which are active only between pH 5.5 and 8.7 and have a molecular weight of about 100,000. Different compounds produce blood damage, skin damage and lethal effects in mice. It is possible to produce antigens to the toxins.

Symptoms Contact between the coral and sensitive skin produces an immediate burning sensation, followed within minutes to hours by red, sometimes itching eruptions. The discomfort and eruptions subside within a day or so and all evidence is usually gone within a week. In very sensitive individuals, pustular lesions or some peeling of the skin may occur. Cases of full thickness skin loss at the burn site have been reported.

Treatment Alcohol has been used to inactivate all remaining nematocysts. This may be followed by generous application of unseasoned meat tenderizer. Papain, the proteolytic enzyme in the meat tenderizer, may neutralize the protein toxin from the nematocysts. Folk medicine has used the freshly squeezed juice of a lime to alleviate the sting. Most experienced divers rub the area and take no further action prior to internal application of lime juice and alcohol at cocktail hour.

Portuguese man-o-war
Physalia physalis

p. 82
Hydrozoa

Toxic element The stings are the result of the injection of venom into the skin by specialized cells called nematocysts found on the tentacles. The nematocysts forcefully expel a hollow, microscopic harpoon, connected to the cell with a hollow thread, into the skin upon contact. The stinging cells have been reported to have sufficient force to penetrate through a rubber glove and enter the underlying skin. The multiple, toxic, proteinaceous compounds are injected into the site of each sting via the hollow attachment. The systemically toxic and hemolytic effects of the venom are caused by the protein physalitoxin, with a molecular weight of 240,000. There are different necrosis-inducing compounds, which are different still from the compounds producing allergic reactions. Antibodies to the venom are produced and may persist for several years. It is curious that the monoclonal antibodies for the lethal factor in the nematocysts also react with the toxin in hornet venom.

Symptoms Contact with the tentacles causes itching, burning wheals in one or more lines, then numbness and pain of the limbs. Severe rigidity of the chest and abdominal muscles is common. The site of the sting may show development of keloids, gangrene, hyperpigmentation and scarring. Systemic effects may include double vision, loss of muscle coordination, spasms of large muscle masses of the trunk, vertigo, fever, vomiting, kidney failure and bladder atony. A 4-year-old girl experienced hemolysis and acute kidney failure after receiving massive stings. Deaths as the result of severe stings have been diagnosed as being the result of the venom acting on the heart.

Treatment It is important to remove any tentacles which remain on the skin to minimize further discharge of stings. Remember that any tentacle may still harbor large numbers of nematocysts capable of discharge and any physically vigorous treatment may discharge additional stings. Some reports state the application of vinegar will inhibit the discharge of nematocysts while the tentacles are being removed. Other authors claim that experimental evidence shows vinegar produces some nematocyst discharge. Ice helps relieve the pain, but in severe cases morphine may be required. Calcium gluconate has been used intravenously to relieve muscle cramps. Potassium appears to counteract the action of the venom on the heart. The calcium blocker Verapamil has been reported to be effective if used early. For anaphylaxis, supportive therapy, including maintenance of an airway, oxygen and fluid supplementation, and application of epinephrine, may be needed.

Notes Sea turtles which feed on jellyfish seem to have an immunity to the toxin in the nematocysts. The man-o-war fish *Nomeus gronovii* seems immune to the stings and stays near *Physalia* seeking protection from predators among the tentacles.

Sea wasp
Carybdea alata

p. 83

Scyphozoa

Toxic element The stings are the result of the injection of venom into the skin by specialized cells called nematocysts found on the tentacles. The nematocysts forcefully expel a hollow, microscopic harpoon, connected to the cell with a hollow thread, into the skin upon contact.

Symptoms Contact of the tentacles with the skin produces intense pain when the nematocysts fire. In cases where the tentacles are smeared about in response to the initial sting, large areas may show redness and blistering similar to burns, with eventual peeling of the skin in affected areas. The toxin may produce muscle contractions due to the release of prostaglandins.

Treatment It is important to remove any tentacles which remain on the skin to minimize any further discharge of stings. Remember that any tentacle may still harbor large numbers of nematocysts capable of discharge. The application of vinegar will inhibit the discharge of nematocysts while the tentacles are being removed. Calcium gluconate has been used intravenously to relieve muscle cramps. The calcium blocker Verapamil has been reported to be effective if used early. For anaphylaxis, supportive therapy, including maintenance of an airway, oxygen and fluid supplementation, and application of epinephrine, may be needed.

Notes The cubomedusae *Chironex fleckeri,* found in Australian waters, is also known locally as a sea wasp, but it is much more virulent: it has been known to kill humans within minutes.

Upside-down jellyfish
Cassiopeia xamachana

p. 84

Scyphozoa

Toxic element   If perturbed by physical contact or sudden turbulence, *Cassiopeia* releases a defensive cloud of nematocysts into the water. On contact, the nematocysts propel a hollow, microscopic tube into the skin of the victim and inject the toxin into the skin through the tube.

Symptoms Contact with a free-floating nematocyst produces a sharp ping of pain followed by a gradually declining itching, burning sensation. Sensitive individuals may show a red swollen mark at the site of the sting for several days.

Treatment Awareness and avoidance of the jellyfish is the best solution. When snorkeling in the presence of these peaceful animals, you should avoid returning to any area disturbed by the turbulence caused by your fins for 15 minutes or more. Thin nylon dive skins or even pantyhose are sufficient to protect the skin from these mild stings. It is also good practice to swim parallel to rather than behind other divers. Hydrocortisone ointments help relieve the usual short-term discomfort.

Sea nettle
Chrysaora quinquecirrha

p. 85
Scyphozoa

Toxic element The stings are the result of the injection of venom into the skin by specialized cells called nematocysts found on the tentacles. In the sea nettle, several different types of nematocysts occur, but the functional difference has not been identified. On contact with the skin, cells forcefully expel a hollow, microscopic thread covered with the viscous venom. The toxins, released in the dermis and the circulatory system, are a group of enzymes and polypeptides which individually may produce anaphylaxis or damage to the heart, nervous system, blood, kidneys or skin. Several tissue-dissolving enzymes are also present. Experimentally, the considerable heart toxicity of one component of the toxin has been demonstrated to be a high-molecular-weight protein. Man may produce antibodies to the venom which persist for many years and cross-react to the venom of other jellyfish.

Symptoms Pain is felt immediately after contact with a tentacle. The sting site may initially be a blanched, itching line on the skin. Within minutes the line becomes red and swollen. In severe envenomations, local muscle cramps may occur. Corneal ulcerations and swelling of the eyelid may result from contact with the eye. After minor contact, the pain and swelling usually disappear within an hour, but may persist or recur in sensitive individuals. The systemic and heart-toxic components of the venom are different from the necrosis-inducing compounds.

Treatment If any tentacles remain on the skin, it is reported that the application of a slurry of baking soda will inhibit further discharge of nematocysts. The application of vinegar causes the discharge of nematocysts. No medication currently available has been shown to be protective or curative against sea nettle stings, but anecdotal reports claim that meat tenderizers with proteolytic enzymes rubbed vigorously into the skin provide symptomatic relief. The proteolytic enzymes may prevent additional discharge of nematocysts, but a mechanism for reversing the influence of injected toxins is unlikely. Experimentally, the use of Verapamil has reduced the systemic effects of the venom in laboratory animals.

Notes This jellyfish has also been known as *Dactylometra quinquecirrha*.

Giant jellyfish

p. 86

Cyanea capillata

Scyphozoa

Toxic element The very fine tentacles commonly break off in turbulent water and drift freely. Thus, a swimmer may contact sting tentacles without seeing any jellyfish nearby. Laboratory experiments have shown that in addition to causing a localized inflammatory reaction, the toxin of this jellyfish is toxic to the heart due to induced electrophysiological changes.

Symptoms Contact with the tentacles produces a sharp, burning pain within 90 seconds. This is followed by a slowly spreading reddened area which usually subsides within 2 days. A highly pigmented area or scar may remain for several months at the source of the sting. The wheals produced by the tentacles are very narrow compared to other jellyfish stings. The rare systemic effects include muscular pain, nausea and drowsiness.

Treatment The symptoms are self-limiting and recovery is seldom influenced by external medications.

Thimble jellyfish

p. 86

Linuche unguiculata

Scyphozoa

Toxic element The nematocysts of the small adult jellyfish may produce a mild sting on sensitive skin, but the larvae produce a more vigorous sting called seabather's eruption.

Symptoms The microscopic larvae are usually induced to sting when they are pressed against the skin. The stings or subsequent rash most often occur under bathing suits. Other occurences are on the feet after wearing swim fins, at the edges of wet suits, and on the chests and abdomen of surfers. The larvae may remain imbedded in clothing and resume stinging after some hours. The intitial stings develop into itchy red swellings or a rash which usually disappears within several days but may last 2 weeks.

Treatment Vigorously bathing with soap and water will eliminate any larvae still on the skin. Clothing should be laundered before being worn again. The skin irritation subsides without treatment, but calamine or other soothing lotions may help supress the symptoms.

Notes While clothing usually protects you from marine stingers, just the opposite is true of this one. It seems that casual contact with human skin in the water does not induce stinging, raher it is only when the larvae are held firmly against the skin that stings result. This stinging phenomenon is sometimes blamed on "sea lice" which are actually arthropod parasites on marine fish and do not produce these symptoms.

Branching anemone
Lebrunea danae

p. 87
Anthozoa

Toxic element All of the anemones have stinging nematocysts, but only a few will penetrate the skin to sufficient depth to produce a sting. Some of the more potent ones will produce blistering on sensitive skin.

Symptoms The symptoms vary from nonexistent to painful stinging of sensitive skin such as that on the inside of the forearm. The site of each sting or tentacle contact is blanched with a halo of red. Various degrees of swelling may be present. In rare severe cases, skin destruction with bleeding and blistering may be followed by ulceration. It is extremely rare for an anemone to produce any symptoms on the thicker skin of the palm of the hand.

Treatment The stings are easily avoided by remembering that anemones are soft but do not like to be stroked. Stings produce an immediate stinging, burning pain which diminishes within minutes or hours. The site of the sting is marked by a red, swollen area which usually disappears within a day.

Black sea urchin
Diadema antillarum

p. 87
Echinodermata

Toxic element *Diadema* have slender, brittle, needle-sharp, microscopically barbed spines. While some published suggestions refer to the spines as carrying toxic substances and other references state the need for surgical removal of the spines, both comments are simple errors with respect to this species. The author and dozens of other divers of his acquaintance have been impaled repeatedly by *Diadema* spines while working or sightseeing underwater. In every case, the irritation was minor and strictly related to the physical damage produced by the spine. The rare cases of infection seem to be related to the introduction of foreign material into the wound by the spine. Some species in this and other genera have the ability to sting and inject venom from the pedicellariae, but none of them occur in the Caribbean.

Symptoms A puncture by a sea urchin spine is similar to that of being impaled by a thorn. A sharp pain is accompanied by regret for one's improvident action! The fragile spine, with its rough, barbed surface, will usually break off in the flesh. A harmless blue or purple pigment will usually be evident at the wound site. In extremely rare circumstances, an individual might show an allergic response to the protein material on the surface of the spine. The epithelial covering of the spine does not generally seem to provoke a foreign protein reaction. More intense pain and short-term swelling are shown by some individuals. Lodging of the more substantial spines of certain other species of urchins in or near joints has resulted in chronic granulomatous inflammatory reactions.

Treatment   Generally, the spine will be absorbed uneventfully by natural body processes, leaving only a transient, residual, purple pigment at the puncture site. Most divers do not treat the injury, but the application of a mild surface antiseptic such as hydrogen peroxide is prudent. The spines are inevitably dissolved by the body's processes within a few days. Attempts at removing or probing for the spine usually result in fragmenting it, introducing bacteria to the wound and further traumatizing minimally damaged tissues. In the rare instances in which a portion of the spine protrudes from the surface of the skin, it may be removed or broken off flush if gently grasped by forceps and pulled straight out. While bacterial infection of the puncture wound might be expected, it is extremely rare. As with any penetrating injury to the body, tetanus prophylaxis should be insured and any unusual or extreme reaction or infection should lead to medical consultation.

Beneficial uses   Perhaps the organic coating or the pigments of the spine have some antibiotic effect, preventing secondary infection of the puncture wounds.

Notes   The roe (ripe ovaries) of several sea urchins are highly esteemed as culinary delicacies in Japan and may appear on Caribbean menus as sea eggs. The short, white-spined sea urchin (*Tripneustes ventricosus*) commonly found on shallow marine grass beds of the Caribbean is one of the sources of this roe, but is reputed to be seasonally toxic.

Cone shells   p. 88
Conus   Mollusca

Toxic element   The yellow, viscous venom of cone shells, produced in a tubular venom gland and stored in a muscular bulb, is injected by a specialized, hollow, barbed tooth called a radula after a prey item is harpooned. Although the venom apparatus is used primarily for feeding, it may be used defensively by driving the radula into the hand of an incautious collector, an octopus or other predator. The most significant effect of the venom is paralysis. Neurotoxic venoms typically interfere with nerve transmission at the synapse between nerve and muscle. Cone toxins have three modes of interference: presynaptically at the nerve end, postsynaptically at the muscle end plate, and in the electrical transmission within the muscle. Chemically, the toxic component of the venom has been identified as a series of peptides composed of 13 to 30 disulfide bonded amino acids. The specific compounds present and their relative proportions vary considerably between species. *Conus imperialis,* which feeds selectively on the marine worm *Eurythoe complantata,* has serotonin as an additional component of its venom.

256

Symptoms The sting produces a sharp, stinging pain followed by a numbness which may spread from the wound site over the entire body and is especially noticeable around the lips and mouth. In severe cases, the inability to swallow or speak, double or blurred vision, excessive salivation and generalized itching may be seen. Paralysis of the muscles may lead to coma and eventual death by respiratory failure. Laboratory experiments have shown that the toxins produce a neuromuscular blockade by binding to the acetylcholine receptors of the muscles. The blockade may be reversed by the anticholinesterase neostigmine.

Treatment Cone stings can be prevented by disturbing the cone and causing it to retract before picking it up. Always handle the cone by the broad part of the shell and do not allow the aperture to rest against your skin. If the tubular proboscis is seen to extend out of the anterior end, the shell should be released immediately.

Notes Stings from *Conus geographus* and *Conus textile* found in the Indo-Pacific Ocean produce a mortality rate in excess of 50%. The lethal dose of this venom is estimated at 1 to 3 micrograms per kilogram, which produces a severe paralysis prior to death. Extracts of the venom ducts have yielded homarine, butyrobetaine, amines, 5-hydroxytriptamine, N-methylpyridinium, lipoproteins and carbohydrates. A peculiar effect of small doses of the toxins is to produce an exaggerated dominant behavior in subordinate lobsters.

Four-eyed octopus

p. 89

Octopus filosus

Mollusca

Toxic element The octopus has a pair of chitinous jaws similar in form to a parrot's beak. The venom is produced in glands near the mouth and enters the body of the victim through the wound produced by the jaws. Each member of the genus produces a toxin which helps immobilize prey species. The blue-ringed octopus in the genus *Hapalochlaena* of Australian waters has a venom which seems to be identical to the tetrodotoxin of puffer fish. Most of the *Octopus* species in the western Atlantic have venom which is particularly toxic to their favorite prey of clams and snails. The cephalotoxin from the salivary glands of *Octopus vulgaris* is composed of several glycoproteins which are paralytic to crabs. There are many other active enzymes such as hyaluronidase and 5-hydroxytryptamine.

Symptoms The bite of the Australian blue-ringed octopus produces severe paralysis to the extent that deaths have been reported due to respiratory failure. Reports of bites in the Caribbean area usually have few significant toxic effects. Numbness or tingling at the wound site accompanied by slight dizziness are not uncommon. One diver has reported a nonpainful swelling of the hand which persisted over a month after an octopus bite; others have reported localized numbness for a week or more. A fishery biologist experienced a sharp pain and a red dot at the bite which turned bluish. Itching at the bite developed after several days and lasted for a month.

Treatment The bite of an octopus from the Caribbean or southwest Atlantic should be disinfected and treated as a minor wound. Any toxic features seem to be localized and transient.

Beneficial uses Octopuses are a major marine food source in many parts of the world. In clear tropical waters, their extremely diverse behavior provides a never-ending source of entertainment for divers.

Notes This octopus has also been known by the name of *O. hummelinki*. Physiologically active compounds identified in the venom of *Octopus* are epinephrine, norepinephrine, serotonin, dopamine, histidine, histamine, tryptophan, polyphenols and others.

Red-tipped fireworm
Chloeia viridis

p. 90

Annelida

Toxic element The sharp bristles, reported to be nonvenomous by some and venomous by others, penetrate the skin and remain embedded.

Symptoms The bristles cause a painful, burning sensation when they penetrate the skin. This is followed by swelling, inflammation and numbness, with a slow-to-heal, irritating wound. Complications, including secondary infection, gangrene and even loss of the extremity, have been reported.

Treatment The bristles can sometimes be removed by the application and removal of adhesive tape at the injured site. Ammonia is reported to relieve the discomfort. The sting site should be carefully disinfected.

Notes This worm has also been known as *Chloeia euglochis*. The bristles of this worm have been included effectively in a virulent internal poison used for criminal purposes.

Orange bristle worm
Eurythoe complanata

p. 91

Annelida

Toxic element The bristles have a hollow, needle-like appearance, with barbs along the shaft that seem to be filled with a fluid. The chemical nature of the bristle contents has not been investigated, but extracts from the body wall show the presence of a strong muscle-relaxing substance. Electron micro-scopic studies of the bristles show them to be hollow with walls of tubules. They have a sharpened tip and no visible glandular connections at the base. Because the preparation of the bristles for microscopic examination requires washing and dehydration, no fluids have been observed in their hollow core. Perhaps the irritating substance is deposited in the bristle at the time of its growth.

Symptoms The bristles penetrate the skin and cause a painful, burning sen-sation, swelling, inflammation and numbness with a slow-to-heal, irritating wound. Small pustules may form within a week or so, then break and release the pus and the bristle. Complications, including secondary infection, gan-grene and even loss of the extremity, have been reported. Systemic cardio-vascular symptoms are reported but are very rare.

Treatment The bristles can sometimes be removed by the application and removal of adhesive tape at the injured site. Ammonia is reported to relieve the discomfort. The sting site should be carefully disinfected.

Green bristle worm, fireworm   p. 92
Hermodice carunculata   Annelida

Toxic element The bristles have a hollow, needle-like appearance and some seem to be filled with a fluid. One investigator reports finding some of the hollow bristles connected to glandular cells at their base. Some investigators claim the bristles are hollow but contain no toxin or glandular organ capable of producing toxin. The chemical nature of the bristle contents has not been investigated. The physiological response certainly far exceeds the minor irri-tation of the equivalent inert spicules of fiberglass when embedded in the skin.

Symptoms The bristles penetrate the skin and cause an immediate painful burning sensation, sometimes followed quickly by a spreading numbness. Some victims show swelling and inflammation with a slow-to-heal, irritating wound. Complications, including secondary infection, gangrene and even loss of the extremity, have been reported.

Treatment The bristles sometimes can be removed by the application and removal of adhesive tape at the injured site. Ammonia is reported to relieve the discomfort. The sting site should be carefully disinfected.

Notes This worm is also commonly called a bearded fireworm.

Honey bee

p. 93

Apis mellifera

Hymenoptera

Toxic element The stinger, located on the tip of the abdomen, is a modified egg-depositing structure which has evolved into a highly effective defense organ. In the case of honey bees, the barbed stinger and associated venom glands and ducts tear lose from the bee after it inserts the stinger into the skin of the victim. The sting continues to pump venom for up to 60 seconds after the bee departs. Honey bees thus sting only once and die as a result of the act. The act of stinging also releases an alarm pheromone which induces other bees to sting. They usually sting only in defense of the colony or when being physically harmed. Foraging bees usually ignore people if they are not molested. The venom is a mixture of biogenic amines, proteins, peptides and enzymes. The major component is a hemolytic polypeptide called melittin, but also present are hyaluronidase, phospholipase A and histamine. The various components are synergistic in producing tissue damage, pain and itching. It is estimated that 500 stings in a short period is lethal to an average adult human, but a single sting has caused death, and individuals have survived receiving over 2,000 stings. The venom of the Africanized bees is no different from that of domestic bees; the colonies are merely more aggressive.

Symptoms A typical sting produces a sharp pain followed by a reddening of the area (often with a blanched center) and a hot swelling. For the majority of the population, the sting will significantly diminish within 24 hours, leaving no adverse after-effects. Some individuals show a systemic reaction, with a rash over the whole body, wheezing, throat constriction, vomiting, abdominal pain and fainting. These symptoms can become so severe as to be life-threatening within 15 to 30 minutes of the sting. Anaphylactic shock reactions may occur within seconds in which difficulty in breathing, vomiting, confusion and falling blood pressure lead to unconsciousness or death. Beekeepers develop a tolerance to bee stings after being stung many times, yet the reaction to stings may become acute suddenly for no obvious reason.

Treatment The sting should be removed from the skin as soon as possible by scraping with a knife blade or fingernail. An attempt to pluck the stinger from the skin will squeeze additional venom into the wound from the venom apparatus. Epinephrine (0.3 to 1.0 ml of 1:1000 concentration) is the treatment of choice for severe systemic allergic response. Calcium gluconate (10 ml of 10% solution) is used if epinephrine is not effective. Steroids have been used if symptoms persist after the above treatments. Oxygen should be provided if respiratory difficulty is observed. In severe cases of swelling of the larynx, a tracheotomy may be required. A person showing anaphylactic response to stings should undergo a desensitization treatment by a physician specializing in allergies.

Beneficial uses The production of honey, beeswax, pollen and royal jelly is a worldwide multibillion-dollar industry carried out in virtually every nation. The associated manufacture of production equipment such as beehives and honey extractors is also a worldwide industry.

Notes Honey bees are responsible for more human deaths in North America every year than any other venomous animal. Individuals who are aware of an extreme sensitivity to bee stings should keep appropriate medication with them at all times. Small syringes filled with epinephrine solution are available as convenient pocket-sized kits. The stings of wasps are sufficiently different from honey bees that a person with a hypersensitivity to one venom may not react unusually to the venom of the other.

North American cow killer
Dasymutilla occidentalis

p. 93
Hymenoptera

Toxic element The females are notorious for having a very painful sting.

Symptoms The sting produces an immediate piercing pain followed by swelling. Normally, both begin to subside within 2 hours, but some persons show an allergic reaction to the venom.

Treatment Ice packs and oral analgesics help relieve the immediate pain. Systemic anaphylactic reactions to the venom may require epinephrine and/or antihistamines.

Notes The wasps of this family (Mutillidae) are called velvet ants because the females are wingless and superficially resemble ants, with vividly colored, velvet-like hair covering the body. These wasps are also commonly known as cow killers, in reference to their vicious and easily provoked sting. The aggressive imported fire ant *Solenopsis invicta* seems unable to penetrate the cuticle of *Dasymutilla,* and other potential predators such as spiders, reptiles, birds and mammals seem to avoid the wasp, either because of the impenetrability of the cuticle or the inevitable retaliatory stings.

Fire ant
Solenopsis geminata

p. 94
Hymenoptera

Toxic element This ant often obtains a firm hold on your skin with the mandibles on the head. Then, using this grip, it arches its back and drives the stinger deeply into your flesh. The ant has a sufficient reserve of venom that it can and will sting several times in succession by pivoting the body and reinserting the sting in a new location. The volume of a single sting varies from about 0.04 to 0.11 microliters. The venom is about 95% alkaloids in the form of 2-alky-6-methylpiperidines called solenopsins, which are insoluble in water and are responsible for much of the pain and pustule formation. The 5% aqueous portion contains at least 4 different non-crossreacting proteins which are responsible for the immunological response of sensitive individuals. Only one of the proteins reacts with the allergens of bee and wasp venoms. Pharmacologically, the venom reduces mitochondrial respiration and uncouples oxidative phosphorylation. These piperidines also block neuromuscular transmission postsynaptically.

Symptoms A localized, intense burning is present at the site of the sting. The venom is injected slowly; thus, the ant is usually still in the act of stinging when discovered. When a nest is disturbed, the ants typically swarm over everything in the vicinity and do not sting until an alarm pheromone is released. It is not uncommon to have many ants on your leg who all sting simultaneously when the signal is given. Fire ants can be brushed off by alert individuals before a full dose of venom is deposited. The normal reaction is to develop an itching wheal within 20 minutes which usually subsides within 2 hours. Small, sterile pustules often develop at the site of the stings from imported fire ants within 24 hours, then heal within a week. Inebriated individuals may select a soft fire ant nest for a nap after a night on the town. There are multiple reports of these individuals receiving as many as 5,000 stings (as evidenced by pustule counts) before being admitted to a hospital. Most show an uneventful recovery from the stings and the hangover, with no systemic involvement. Certain individuals show a large local reaction, with redness and associated swelling extending to the entire limb before the symptoms begin to subside after 48 hours. Sensitized individuals receiving only a few stings may show a systemic anaphylactic reaction, with a general rash, body swelling, labored breathing, rapid heartbeat, visual disturbances, sweating and reduced blood pressure. As with other hymenoptera stings, this indicates a need for prompt medical treatment..

Treatment Washing the site removes the surface venom, and the discomfort will usually subside within several hours. Treatment of the stings with antibiotic, antihistamine or fluorinated hydrocortisone had no influence on the course of the pustules in controlled experiments on volunteers. After the pustules burst, prophylactic treatment with local antiseptics reduced the incidence of secondary infection. Oral antihistamines were valueless in treating the stings. The formation of the pustule and its development are unrelated to the systemic anaphylactic response shown by some patients. The treatment recommended for large local reactions has included oral corticosteroids, oral H_1 and H_2 antagonists and analgesics. In cases of multiple stings or in allergically sensitized individuals, immediate medical assistance in the form of epinephrine and supportive therapy may be needed. Swelling of the larynx may require an emergency tracheotomy if a hospital setting with endotracheal tubes is not immediately available. Injected epinephrine usually relieves systemic symptoms rapidly. Intravenous calcium has been recommended in severe cases of toxicity. Immunotherapy for systemic response to fire ant venom shows promise of being a safe and effective prophylactic treatment.

Beneficial uses Solenamine, the primary toxin in the venom, has selective antifungal and antibacterial properties as well as being insecticidal.

Notes Two similar fire ants have been introduced near Mobile, Alabama, this century. The black fire ant *S. richteri* was introduced around the turn of the century from Argentina and the red fire ant *S. invicta* around 1930 from Brazil. Both have been broadly spread through the Southeast. They build-mounds up to 18 inches in diameter by 10 inches high, produce a more severe sting and may swarm aggressively up your leg if you should inattentively stand on or near a disturbed nest.

Wasp, hornet, yellow jacket
Polistes, Vespula, Dolichovespula

p. 95
Hymenoptera

Toxic element The barbed stinger of hornets and wasps is a modified ovipositor composed of a pair of barbed lancets which are inserted into the victim with an alternate sawing motion pushed with the strength of the entire body. With experience, the gripping of the legs can be detected by the victim before the sting is sufficiently inserted to inject venom. The venom duct injects venom between the lancets under the surface of the skin. The venom is a complex mixture of compounds including histamine, serotonin, noradrenalin and dopamine. There are several proteolytic and fibrolytic enzymes, including hyaluronidase and phospholipases A and B, which break down the victim's tissues and facilitate the spread of venom. Sensitivity to the venom of one species of bee or wasp does not necessarily imply the equivalent sensitivity to all stinging hymenoptera because each has a unique combination of antigens.

Symptoms Depending on the location, the sensitivity of the individual and the severity of the sting, the pain may range from uncomfortable to agonizing. The immediate localized swelling of variable magnitude, followed by a reddening of the area, are distinctive and subside in a day or so. An increase in capillary permeability, dilation of blood vessels and a contraction of smooth muscles all contribute to the symptoms. Certain individuals develop an allergic sensitivity to the stings which starts with a generalized itching and proceeds with a rash over the whole body, wheezing, throat constriction, vomiting, abdominal pain and fainting 30 minutes after a sting. Future stings may produce a generalized swelling of the body and swelling of the throat, which produces difficulty in breathing and swallowing. If not promptly treated, death due to asphyxiation may occur. Anaphylactic shock reactions may occur within seconds of a sting. Difficulty in breathing, vomiting, confusion and falling blood pressure may lead to unconsciousness or death. Although sting symptoms typically have a rapid onset and recovery, deaths have been recorded as long as 4 days after a sting.

Treatment There are several commercially produced suction devices which may extract some of the venom if applied immediately. For only a few stings, a nonprescription antihistamine tablet may help reduce swelling. Cold packs help relieve the pain. In cases of multiple stings, or in allergically sensitized individuals, immediate medical assistance in the form of epinephrine, antihistamines and supportive therapy may be needed. Intravenous calcium has been recommended in severe cases of toxicity.

Notes The paper for the nests is made from wood fibers and has a remarkable resemblance to the commercial paper made by man from the same source. One of the protein toxins in the venom of the wasp which affects the central nervous system has been found to cross-react to the same monoclonal antibodies that react to the lethal factor in the nematocysts of the man-o-war jellyfish *Physalia physalis*, indicating a significant similarity of chemicals. The purified compound has a molecular weight of about 43,000.

Yellow fever mosquito   p. 97
Aedes aegypti   Diptera

Toxic element The bite of the mosquito is actually a puncture as the mouth parts penetrate the skin seeking a blood meal. The itching response to the mosquito's feeding is an allergic reaction to the saliva, which is injected to aid in the flow of blood. The saliva contains anticoagulants, haemolysins and other enzymes which assist the mosquito in rapidly obtaining a meal before being swatted. The saliva also contains substances which reduce the pain of the initial bite, but unfortunately contribute to later skin reactions.

Symptoms The brief itching pain of the bite passes within minutes. For those not acclimated to mosquito bites, the itching may continue for some time. For those only rarely bitten, the response may be a red, itching wheal which lasts several days.

Treatment For areas with extremely high densities of mosquitoes, covering the body with clothing and a headnet is the only feasible prevention. With low and modest densities, commercial insect repellents are very effective at preventing feeding. The application of proprietary lotions containing benzocaine relieves the continuing itch of sensitized individuals, but has no effect on the duration of the itching wheals.

Beneficial uses   It has been suggested that regular and persistent attacks by feeding mosquitos might contribute sufficient anticoagulants to inhibit coronary thrombosis.

Notes   Their blood feeding habit makes them the most important vector of the human diseases yellow fever and dengue fever. Yellow fever is still present in less-developed countries in the tropics, but it has now been greatly controlled by the vaccination of susceptible individuals. Dengue fever continues to be of public health concern in the rainy season in many of the areas of the Caribbean.

Malaria mosquito   p. 98
Anopheles quadrimaculatus   Diptera

Toxic element   The bite of the mosquito is actually a puncture as the mouth parts penetrate the skin seeking a blood meal. The itching response to the mosquito's feeding is an allergic reaction to the saliva, which is injected to aid in the flow of blood. The saliva contains anticoagulants, haemolysins and other enzymes which assist the mosquito in rapidly obtaining a meal before being swatted. The saliva also contains substances which reduce the pain of the initial bite, but unfortunately contribute to later skin reactions.

Symptoms   The brief itching pain of the bite passes within minutes. For those not acclimated to mosquito bites, the itching may continue for some time. For those only rarely bitten, the response may be an itching wheal which lasts several days.

Treatment   For areas with extremely high densities of mosquitoes, covering the body with clothing and a headnet is the only feasible prevention. With low and modest densities, commercial insect repellents are very effective at preventing feeding. The application of proprietary lotions containing benzocaine relieves the continuing itch of sensitized individuals, but has no effect on the duration of the itching wheals.

Notes   This mosquito is most notorious as being part of the life cycle of the protozoan red blood cell parasite *Plasmodium,* which causes malaria. A single bite from an infested mosquito is sufficient to infect a person with malaria, subjecting that person to outbreaks of the disease for the rest of his or her life. There are several drugs on the market which can be taken as a prophylaxis when traveling in malaria areas. Modern medicine is now able to cure malaria, but drug-resistant strains are becoming more common. The oral ingestion of large amounts of thiamin hydrochloride seems to reduce the palatability of humans to mosquitos. Insect repellents which contain deet (N,N-diethyl-m-toluamide) seem to be the most effective in reducing the number of bites. Clothing treated with permethrin also provides almost absolute protection of the covered area from bites. Avon Skin-So-Soft provides very short-duration protection from mosquito bites. The saliva from this mosquito prevents the clotting of blood, even at a dilution of 1 to 10,000, and it has been suggested that regular exposure to the bites of this mosquito might reduce or prevent coronary thrombosis.

House mosquito
Culex pipiens

p. 99
Diptera

Toxic element The bite of the mosquito is actually a puncture as the mouth parts penetrate the skin seeking a blood meal. The itching response to the mosquito's feeding is an allergic reaction to the saliva, which is injected to aid in the flow of blood. The saliva contains anticoagulants, haemolysins and other enzymes which assist the mosquito in rapidly obtaining a meal before being swatted. The saliva also contains substances which reduce the pain of the initial bite, but unfortunately contribute to later skin reactions.

Symptoms The brief itching pain of the bite passes within minutes. For those not acclimated to mosquito bites, the itching may continue for some time. For those only rarely bitten, the response may be an itching wheal which lasts several days.

Treatment For areas with extremely high densities of mosquitos, covering the body with clothing and a headnet is the only feasible prevention. With low and modest densities, commercial insect repellents are very effective at preventing feeding. The application of proprietary lotions containing benzocaine relieves the continuing itch of sensitized individuals, but has no effect on the duration of the itching wheals.

Notes The females serve as intermediate hosts of filariasis (elephantiasis) and encephalitis in humans; heartworms in dogs; malaria and fowl pox in birds; and encephalitis in horses.

Sand flies
Culicoides furens

p. 99
Diptera

Toxic element The saliva injected into the skin during the feeding of this gnat produces itching as an allergic reaction. The filarial worm *Mansonella ozzardi* is transmitted by the bites of this virulent pest.

Symptoms A stinging itch is the first response to a bite. Sensitive individuals or those unaccustomed to being bitten will often develop a swollen red welt which will last several days.

Treatment Considerable time, effort and expense have been applied to developing chemical repellents to protect mankind from the feeding of these irritating little insects. The common commercially available products containing deet (N,N-diethyl-m-toluamide) are the most effective. Several other products, such as Avon's Skin-So-Soft, Claubo and Johnson's baby oil, also help prevent bites because their oiliness traps the insects on the skin surface, not because they have any repellent action.

Notes These pesky little bloodsuckers are also known as no-see-ums, punkies and a variety of other profane appellations.

Biting spider   p. 100
Chiricanthium   Arachnida

Toxic element The venom has not yet been fully analyzed.

Symptoms Sudden, sharp pain is felt at the site of the bite, followed by local aching or tingling and itching. The pain and tingling may be severe as they increase over the first 30 minutes to an hour and they may extend through the limb to the adjacent trunk. A red wheal appears immediately after the bite, then becomes tender to touch and pressure as the skin temperature is elevated. The patient is often restless. Blood values and vital signs remain normal, but abdominal cramping has been reported. The intense pain diminishes after about 4 hours, but residual pain may be present for several days. Mild neurotoxic symptoms may be felt. A necrotic lesion to 3 cm (1 in) with a raised margin, swollen surroundings and depressed center often develops. Complete healing of the bite may occur as soon as 10 days, but severe lesions may last as long as 8 weeks.

Treatment Symptomatic treatment with oral antihistamines and/or topical antipruritics is usually adequate to relieve discomfort and anxiety. Intramuscular meperidine hydrochloride and dexamethasone injections have been reported as relieving the intense pain in severe cases. Supportive therapy may be needed.

Notes These spiders are also known as running spiders. The *Chiricanthium* in the Caribbean has been reported as being the species *inclusum*.

Black widow spider   p. 101
Latrodectus mactans   Arachnida

Toxic element The venom of the black widow is produced in glands in the cephalothorax and is injected through the hollow fangs by muscular contractions at the time of bite. It is made of several components which are variations of the neurotoxic protein latrotoxin. Alpha latrotoxin, most active in man, with a molecular weight of about 130,000, disturbs the flow of calcium across membranes and destroys nerve terminals, preventing nerve impulses from reaching muscles. The venom is more toxic than most snake venoms, with an LD_{50} of 0.9 mg/kg in a mouse. The venom of the "brown widow," *L. geometricus,* found in south Florida and worldwide in the tropics, is 3 times as potent, but few bites are known because the spider is less irritable and produces less venom. The injection of venom is controlled by the spider; thus, a bite from the same spider may range from inconsequential to severe.

Symptoms The bite is felt as a sharp pin-prick, but may be so minor that it is not noticed. Bites in which the spider injects little venom develop no further symptoms. A serious bite is often followed in 5 to 60 minutes by a dull pain which involves the entire limb. Muscular cramps begin within 15 minutes to several hours after the bite. There is usually little evidence at the bite site, but the area around the puncture wounds may be blanched with redness surrounding it. Sweating and twitching of the muscles are common. Pain and the feeling of pressure in the chest, abdomen and lumbar region, along with weakness of the legs, are typical. The patient is often anxious and restless, with labored breathing and a cold, clammy skin. Rigidity of the abdominal and intercostal muscles is present in most serious cases. Muscle cramps and intense abdominal pain sometimes lead to shock and delirium. Rapid heart rate and high blood pressure often peak at dangerous levels from the second to fourth hours. Later, the heart rate may be abnormally slowed. In most untreated cases, symptoms will begin to decline after 48 hours and resolve within a week. Significant weight loss is often observed. Weakness, malaise and muscle pain may persist for many weeks. Deaths from the bite of this spider have been reported regularly and seem to average about 4% to 6% of reported bites. In a recent 10-year period in the U.S., about 14% of all deaths due to poisonous and venomous creatures were due to black widow bites.

Treatment The best preventive measure is an alert awareness of the possibility of spiders' presence when working in an area which has not been disturbed for some time. In the event of a bite, it is useful to capture the spider to enable positive identification. In the absence of the spider, clinical diagnosis is based on pain in a localized area accompanied by a rigid, non-tender abdomen and other factors. Intravenous calcium gluconate (10 ml of 10%) has helped relieve the muscle cramps and pain and may be repeated as needed. Methocarbamol given initially intravenously and later orally has been found useful by some and without value by others. Local treatment of the bite is of no value and nothing is gained by attempts to remove the venom by incision and suction. Corticosteroids have given variable results. Meperidine hydrochloride or morphine is effective in relief of the pain.

In severe bites of very young, old, pregnant or hypertensive patients, antivenom may be needed. The venom is prepared from the serum of immunized horses, so all the normal precautions for the use of horse-serum products should be taken. (See appendix on antivenom.) The antivenom may be given intramuscularly, or intravenously diluted 1 to 10 in sterile saline and administered over a 15-minute period. One vail usually provides adequate treatment, but if symptoms do not subside, a second dose may be necessary. Provisions should be available in case the need arises to treat acute anaphylaxis resulting from horse serum administration. In the U.S., contrary to most other countries, antivenom is administered only in severe cases, as the risk of anaphylaxis due to the horse serum is considered greater than the relief of suffering from the symptoms. A specific antivenom for *Latrodactus mactans* venom is available from Merck, Sharp and Dohme, a pharmaceutical company in Rahway, New Jersey.

Beneficial uses The venom is used as a tool in the scientific investigation of nerve and membrane physiology. The venom has been found effective in accelerating recovery from botulism-induced paralysis in laboratory experiments. The venom destroys the nonfunctioning nerve material, inducing regeneration of new, active nerve endings.

Notes The genus is derived from the Greek *latro,* meaning "robber" and *dect,* meaning "bite." The species is from the Latin word meaning "murderous." This spider is also known as the hourglass or shoebutton spider. Certain Native American tribes used the venom to poison arrows.

Brown recluse spider
Loxosceles reclusa

p. 102
Arachnida

Toxic element The venom is about 26% protein and contains a complex mixture of tissue-destroying enzymes, including collagenases, proteases, phospholipidases and hyaluronidase. The skin-necrotizing factor has been identified as sphingomyelinase D and has a molecular weight of about 24,000. The venom also has components producing a breakdown of blood cells and platelet aggregation. The mechanism of primary tissue damage and localized vascular occlusion by thrombosis are still matters of debate.

Symptoms The bite is mild and feels about like an ant sting. Within 30 minutes, the area has a burning, itching sensation and the area of the fang punctures is blanched and surrounded by a red ring. The red area enlarges for about another 8 hours and may become irregular in shape with small skin hemorrhages. The fully developed lesion has a central white vesicle with an extensive swollen and red area which may be edged with a blue-white border. A local necrosis, particularly severe over a layer of fat, often develops. After the initial scab falls off, an open ulcer with undermined edges may remain at the site. Without further treatment, the ulcer may persist and grow for weeks or months and penetrate the underlying muscle tissue. Systemic effects may include chills and sweats, stomach cramps, vomiting, jaundice, enlarged spleen, breakdown of blood cells, blood in the urine and kidney failure. Breakdown of blood cells leading to kidney failure has resulted in deaths. The bite of this spider should not be treated at home; the patient should be transported to a major hospital. In tropical areas, the diagnosis should carefully differentiate cutaneous leishmaniasis.

Treatment The confirmed identification of the species of the offending spider is an important component of the treatment plan. Dapsone and brown recluse antivenoms, given soon after the bite separately or combined, have been about equally successful in producing a successful outcome. Corticosteroids have also been recommended for immediate treatment and in patients showing systemic symptoms, but do not seem to provide a consistent, reliable benefit.. The activity of sphingomyelinase is greatly reduced below 37°C, indicating that ice packs may delay or reduce necrosis if applied soon after the bite. Hyperbaric oxygen at two atmospheres for 60 to 90 minutes twice a day is reported to produce significant improvement in the necrotic lesions. Systemic support may be needed for hemolysis. Skin grafts to repair extensive necrotic lesions are seldom effective until at least 8 weeks after the bite.

Chigger   p. 103
Eutrombicula alfreddugesi   Arachnida

Toxic element The toxin is actually an allergic reaction to the saliva of the larvae. The larvae feed on lymph and skin tissues partially digested by salivary fluids injected by the nymph. As the larvae feed, the host tissue coagulates and forms a tube around the feeding area which extends deeper as feeding proceeds. The larvae do not feed on blood.

Symptoms Most bites occur on the ankles or calves or where clothing is constricting. The intense itching begins 3 to 6 hours after the chigger has attached and may persist for over 2 weeks. The associated swellings may take hours or even a day to fully develop. Bites tend to be more frequent in places where clothing is tight. Individuals develop a tolerance and fail to react to bites if regularly bitten. Chiggers cause swelling, discomfort and sometimes abscess on birds. Heavily parasitized young chickens and turkeys may become depressed and refuse to feed and eventually die of starvation and exhaustion. Chiggers do not cause swelling in reptiles; a single box turtle may be infested with as many as 500 larvae.

Treatment Commercially available insect repellents such as deet, applied to the skin, socks and shoes, and openings in clothing, afford good protection against bites. A good field habit in chigger areas is to never sit on the ground. Benzocaine solutions, phenol or menthol in calamine lotion, or topical corticosteroids have been suggested to help relieve the intense itching. One lotion particularly effective in relieving the itch is composed of benzocaine 5%, methyl salicylate 2%, salicylic acid 0.5%, ethyl alcohol 73% and water 19.5%. Secondary infection should be given prompt antibiotic therapy.

Notes The genus is also known as *Trombicula*. These mites are also called redbugs, red mites and harvest mites in English-speaking areas and *tlalzahuatl*, *coloradillo* and *bicho colorado* in Spanish-speaking areas.

American dog tick
Dermacentor variabilis

p. 103
Arachnida

Toxic element Ticks of 43 species in 10 genera have been found to cause tick paralysis by various mechanisms in mammals and birds. The paralysis is caused by a neurotoxin in the saliva of the pregnant female tick after it has been feeding for 5 to 7 days. The symptoms seem worse if the tick is attached on the neck or spinal column. Extensive experimentation has shown the pathology to be an impairment of the transmission of motor nerve impulses. It still seems a mystery why an animal may harbor hundreds of ticks with no symptoms while another may become symptomatic due to the feeding of a single tick.

Symptoms It should be remembered that tick bite is common, but tick paralysis is rare. The preliminary symptoms of tick paralysis are a feeling of malaise and irritability with a loss of energy and appetite, which may be followed by violent headaches and severe vomiting. Tingling and numbness may be present. The first signs of paralysis show in the lower limbs, with weakness and loss of coordination. The paralysis ascends to influence all the limbs and produces difficulties in speech, breathing, chewing and swallowing, accompanied by extreme apathy. The senses are reduced and reflexes are impeded. There is no fever or pain and blood values are normal. A single female tick can paralyze and kill an adult human. The ultimate cause of death in both humans and dogs is respiratory paralysis.

Treatment Removal of the tick results in recovery from the disease within hours or days. Ticks usually attach to feed on hairy parts of the body and may be difficult to find, but if the victim has been exposed to ticks, a diligent search is in order. Ticks may be prevented from biting and attaching to the skin by various commercial formulations of deet, or more effectively by the treatment of clothing with various compounds such as Permanone (permethrin). Numerous methods of removal of ticks with mouth parts imbedded in the skin are recounted in folklore. A tick grasped firmly with forceps and steadily pulled upward generally will be removed intact. A nail warmed in a match flame, then applied to a tick's back until the legs wiggle, allows the tick to be gently extracted. Coatings such as oils or fingernail polish, which smother the tick by clogging its breathing pores, also cause the tick to release the grip of its mouth parts.

Disease transmission Ticks are second only to mosquitoes as important reservoirs and vectors of many viral, bacterial and protozoal diseases. *Dermacentor variabilis* is the principal vector of the rickettsial disease Rocky Mountain Spotted Fever. It is also a vector of the bacterial disease tularemia and can pass the disease on to its offspring through the egg. Lyme disease caused by the spirochaete bacteria *Borrelia burgdorferi* was recognized as a human health problem in 1975 and is now the most commonly reported tick-borne disease in North America. The bacteria is spread by the bite of the black-legged tick *Ixodes scapularis,* whose adult host is commonly field mice or white-tailed deer. Many wild mammals are hosts for the tick and reservoirs for the disease.

Scorpion   p. 104
Centruroides   Arachnida

Toxic element Scorpions sting by thrusting the aculeus, or stinger, forward over the back. The tapered aculeus generates a flaring puncture wound with venom deposited deep in the wound by the tip. Upon withdrawal of the aculeus, the host tissues elastically contract and often squeeze some of the venom out onto the skin surface. The venom is a complex mixture of more than 10 protein components and several additional nonprotein factions. The long-chain peptides affect sodium channels, while short-chain peptides block potassium channels. Some of the toxic components are very active against insects, while other primarily affect mammals. The toxicity of scorpion venom to mammals varies over more than a hundredfold range depending on the species of scorpion and method of administration. *Centruroides noxius* is one of the most toxic, with an LD_{50} I.P. of 0.26 mg/kg in the mouse. While the most lethal scorpions in the New World belong to the genus *Centruroides*, the stings of the species of that genus found in Florida and the Caribbean are not generally life-threatening.

Symptoms An instantaneous burning pain is sometimes accompanied by redness and swelling. Scorpions squeezed inside a pant-leg rapidly whip the tail, resulting in the multiple stings "like a red-hot needle." Symptoms generally subside greatly within 2 hours. In particularly sensitive individuals, sweating, pallor, and increased respiratory and heart rates with elevated blood pressure may be present.

Treatment There are no first-aid measures which significantly alter the course of minor scorpion sting. Immobilization and cooling of the afflicted part helps reduce the pain. Barbiturates or chlorpromazine may help relieve the intense anxiety often accompanying scorpion sting. Morphine or Demerol given to treat the pain synergize the toxicity of the venom of scorpions in this genus and should be avoided as a scorpion sting treatment. For very sensitive individuals showing systemic symptoms, propanolol hydrochloride, calcium gluconate and hydrocortisone have all been used effectively. Symptoms rarely require oxygen or respiratory assistance. Intravenous calcium and antivenom are seldom needed for the scorpions in Florida and the Caribbean.

Beneficial uses The polypeptide active compounds in scorpion venoms have aroused intense medical interest because of their potential use as tools in neurobiological research.

Notes The ancient Greeks prescribed riding backwards on a donkey as a sure cure for scorpion sting. Since then, the variety of treatments used both by folk medicine and physicians is truly astonishing. None of the treatments other than the specific antivenom for that species of scorpion has shown any clinical effectiveness. Almost all scorpions are fluorescent when illuminated with long-wavelength ultraviolet light. Professional biologists use portable ultraviolet lights at night to find foraging scorpions. Searching around a house and its environs using this technique at night will allow the capture and removal of many scorpions.

Centipede
Scolopendra subspinipes

p. 105
Myriapoda

Toxic element When centipedes bite, they impale the victim and inject venom from a large pair of hollow poison fangs located near the mouth. The two fangs work in opposition to each other, rather like ice tongs. The venom contains histamine, 5-hydroxytryptamine and various digestive enzymes. It has an LD_{50} of 0.75 mg/kg in laboratory mice. A specific protein toxin, with a molecular weight of 60,000, isolated from the venom has an LD_{50} of 41 micrograms/kg in mice. The development of a progressive immunity to centipede bites has been reported.

Symptoms   In spite of its spectacular appearance and occasional aggressive behavior, this centipede is not as bad as it looks. Bites produce significant immediate pain followed by swelling and tenderness that usually diminish within a few hours. In severe bites, the swelling and pain may last for a week or more and may be followed by local necrosis. There is only one recorded fatality from the bite of this centipede.

Treatment   Ice packs help alleviate the initial pain for many people. The injection of a local anesthetic at the bite site produces prompt relief from the pain and considerable psychological reassurance to a person bitten by such a ferocious-looking creature.

Beneficial uses   A decoction of the powdered, dried body is used in Chinese folk medicine to treat tetanus and convulsions. The body is ground with tea leaves and made into a paste for external applications to ulcers and other skin conditions.

Millipede   p. 106
Rhinocricis arboreus   Myriapoda

Toxic element   Many of the segments contain paired repugnatorial glands whose ducts are aimed upward and to the side. When irritated, the millipede can squirt a toxic yellow mist composed of p-benzoquinones 80 cm (33 in) or more into the air. Alternatively, the millipede can cause the toxins to ooze onto the surface of each segment. Typical behavior is for the millipede to ooze if lightly disturbed, to spray when more strongly disturbed and to spray from several segments when vigorously jostled. Only a small amount of the available toxin is used on each shot, so the millipede can continue to produce bursts of toxic mist with repeated threats.

Symptoms   The toxic mist produces an intense, burning pain in the eye, leading to a swelling which may persist for several days or more. Breathing the mist produces a burning in the upper respiratory system, coughing and a transient shortness of breath. The toxin on the skin produces dark-brown stains which may form into blisters. After the blisters heal, the site of the burn may still be evident after more than a year. Poultry are sometimes rendered permanently blind after pecking at these millipedes.

Treatment   The eyes should be copiously rinsed with clean water as soon as possible after being fogged. Local anesthetic drops may be needed to reduce the pain. Cold compresses help alleviate the pain. The secretion should be washed from the skin with soap and water. If alcohol is available, the benzoquinones should be swabbed from the skin with this solvent.

Notes The toxic fog renders these millipedes almost invulnerable to predators, but toads and box turtles have been seen to eat them with no visible adverse effects.

Stingray   p. 107
Dasyatis americana   Fish

Toxic element The dorsal surface of the tail of the stingray has one or more bony barbs with sharp hooked serrations on the edges. The barb is used defensively only when the animal is stepped on or otherwise physically molested at which point the tail is whipped upward and forward driving the barb into the flesh of the offender. The venom is released from the 2 lateral grooves in the barb when the integumentary covering of the barb is torn as it penetrates the victim. Withdrawal of the barb often leaves a severe laceration and may leave parts of the integumentary sheath or even part of the barb in the wound. The venom is a toxic protein with a molecular weight exceeding 100,000 which is rapidly inactivated by heating. The LD_{50} is about 2.9 mg/kg of the protein IV.

Symptoms The grievous nature of stingrays was known and reported by the ancient Greeks, who noted that they caused "gangrene of the wounded flesh." Penetration of the skin by the barb on the tail produces intense local pain and swelling which increase over a period of 90 minutes and are considerably out of proportion to the extent of the wound. Vomiting, muscle cramps, abdominal pain, weak dizziness, rapid heart rate, sweating, and shock with fall of blood pressure may be present. Discoloration may extend several centimeters from the wound within 2 hours. Deaths have resulted from penetration of the spine into the abdomen or chest.

Treatment Fragments of the integumentary sheath should be removed as soon as possible and the wound irrigated to remove any residual venom and sheath fragments. The toxin may be partially destroyed by soaking the extremity in water as hot as tolerable (about 113°F or 45°C) for 30 to 90 minutes. Hot water also helps alleviate the pain. In the absence of heat treatment tissue necrosis with abscess or chronic draining open wounds requiring extended care are common. The affected part should then be elevated. Analgesics such as meperidine hydrochloride help relieve the severe pain. The wound should be cleaned of all foreign material before closing. This may require exploratory surgery if the barb has pierced the pleura or peritoneum. Prognosis is poor in the presence of an unremoved integumentary sheath within the abdominal or chest cavity. Oxygen and supportive care for shock should be available. In severe stings, cardiac and respiratory functions may be impaired by the venom and should be monitored and maintained as needed. Convulsions have been reported and deaths have occurred subsequent to abdominal stings. Tetanus prophylaxis and systemic antibiotics are prudent measures. Folk medicine has used the leaves of the plant *Astronium graveolens* commonly called glassy wood or Zebra as a poultice to relieve the symptoms of the sting.

Beneficial uses Native Americans used the tail spines on the tips of arrows to produce particularly vicious wounds.

Notes Shuffling the feet while wading reduces the probability of stepping on stingrays or stonefish.

Scorpion fish   p. 108
Scorpaena plumieri   Fish

Toxic element The dorsal pelvic and anal spines have an associated venom gland. If allowed to penetrate the skin the spines deposit a venom which causes intense pain and swelling. The pain and swelling may extend to the entire limb. The venom is composed of highly complex heat-sensitive proteins with a molecular weight in the range of 60,000. The total unrefined venom has an LD_{50} of about 2.6 mg/kg in mice.

Symptoms The sting immediately causes an intense throbbing pain. Victims may become hyperactive and writhe around on the ground in pain. The site of the sting is blanched at first then assumes a dark red or bluish color. A firm red swelling develops within a short period and may extend through the entire limb. Pain of great intensity may continue for several hours. Perspiration, pallor, vomiting, diarrhea, rapid heartbeat, convulsions and loss of consciousness are all common symptoms. Electrocardiograms of victims show wandering of the pacemaker, prolongation of the PQ interval and other changes in heart function. Muscular aches and shortness of breath may be present for several weeks after other symptoms have resolved. Adverse effects on the general health may be present for several months.

Treatment The toxin may be partially destroyed by soaking the sting in water as hot as bearable (about 113°F or 45°C) for 30 to 60 minutes. Repeated dunking of the extremity in water too warm to allow sustained immersion may be even more effective but the limb must be periodically removed or blistering is possible. The wound should be irrigated as thoroughly as possible to remove residual venom. Local infiltration of the wound site by analgesics and or intravenous analgesics may be needed to relieve the severe pain. Standard procedures for the prevention and treatment of shock should be instituted. Secondary shock is possible due to the action of the venom on the cardiovascular system. Bleeding should be encouraged and the wound should

276

be cleaned of all foreign material. Tetanus prophylaxis and systemic antibiotics are prudent measures. An anativenom is commercially produced to counteract the dreadfully painful stings of the stonefish *Synanceia*. The stonefish is another member of the same family which has caused human deaths when the sting is untreated. The stonefish is found in the shallow warm waters of the Central Pacific west to the coast of Africa and north to the Red Sea and Gulf of Oman.

Beneficial uses Scorpion fish are highly esteemed as food fish.

Notes Another member of this genus *S. scrofa* is one of the major ingredients in the classic Mediterranean fish soup bouillabaisse.

Atlantic puffer, Porcupine fish   p. 109
Sphoeroides maculatus, Diodontidae   Fish

Toxic element Fish which are toxic due to their tetrodotoxin content have been known by the Chinese for 5000 years. The toxin is known chemically as aminoperhydroquinazoline and is found most notoriously in puffer fish of the family Tetraodontidae. These are known and served as fugu in Japan. The toxin produces a neuromuscular block by acting selectively to reduce sodium permeability of membranes without affecting potassium movement. It is also commonly found in ocean sunfishes (*Mola mola*) and porcupine fishes of the family Diodontidae. The toxin is most concentrated in the reproductive organs, eggs, liver and intestines with lesser and variable amounts in the skin. The toxin is water-soluble but is not inactivated by cooking. The concentration of toxin in the reproductive organs and the total amount of toxin in the fish increase with the approach of the spawning season. The major muscles of the body are free of the toxin; thus a carefully cleaned and skinned fish is edible without any adverse effects to the extent that entire gourmet restaurants have developed a thriving trade based on serving fugu. I find the flesh to be delicious and of excellent texture. On the other hand tetrodotoxin is one of the most toxic nonprotein substances known with an LD_{50} of 12 micrograms/kg I.P. The molecule bears no structural relationship to any other known natural compound.

Symptoms The symptoms usually appear within 5 to 30 minutes of eating the toxic fish. The victim usually feels weakness or numbness or both accompanied by a tingling of the lips tongue and throat. Pallor sweating salivation and a weak rapid heartbeat with reduced blood pressure may be evident. As the symptoms progress the heart shows arrhythmias and is slowed. The pupils are initially constricted but as the disease progresses they dilate with loss of the pupillary and corneal reflexes. A disturbance of proprioception may occur in which the relative weight of objects cannot be determined.

277

Muscular paralysis may be present without the loss of deep tendon reflexes. Speech may be impeded or absent due to paralysis of the vocal cords. A numbness and a feeling of floating may progress to general paralysis and respiratory arrest within 6 to 24 hours of eating the toxic fish (mortality has been reported as promptly as 17 minutes after eating toxic fish). Death may ensue within 30 minutes of eating a large dose of the toxin such as occurs when eating the liver. Mortality in severe cases of toxicity may exceed 50%. Although completely immobilized by paralysis and taken for dead, victims have recovered and report that their mental faculties remained acute.

Treatment There is no antidote for tetrodotoxin. Treatment is supportive and directed at elimination of residual toxin from the digestive system. An emetic should be given if vomiting has not taken place. Lavage with 2% sodium bicarbonate within the first hour may help neutralize the toxin in the stomach. Activated charcoal is effective in binding the toxin and preventing further gastrointestinal absorption. Laxatives and enema may help remove the toxin. Ventilatory assistance or control with supplemental oxygen increases chances for survival. Many patients seemingly near death can be saved by providing vigorous respiratory support. As the toxin acts as a vasodilator, fluids to maintain circulation and kidney function may be needed. The toxin accumulates in the kidneys and inhibits urine secretion; thus diuretics with central stimulatory action may be advised. Several pharmaceutical treatments have been recommended but no controlled clinical trials have been conducted to confirm their efficacy. Antihistamines and corticosteroids have not been useful. Diuretics with central stimulatory action help overcome the inhibition of urinary secretion. Tetrodotoxin does not produce liver dysfunction; thus liver activating drugs have not been found to be valuable. Survival for 24 hours provides an excellent prognosis for full recovery.

Beneficial uses Tetodotoxin has been used (in very small dosages) in Oriental medicine to treat arthritis, asthma, headaches, impotence, itching and tetanus. It is deemed particularly useful as a treatment for neuralgia and convulsions and as a muscle relaxant. A recent compilation of clinical uses has shown the toxin to be effective in relieving several different sources of chronic pain. Tetrodotoxin is used as a tool in modern medical research on the excitation of muscle and nerve fibers. The liver is reputed to contain 15 times as much vitamin D as cod liver oil but as this is the epicenter of toxin accumulation it would seem imprudent to eat it. The Atlantic puffer has been marketed commercially as a food fish both fresh and frozen on the east coast of the U.S. Where available it is enthusiastically received by consumers. The American market has not required the compulsory licensing of trained fugu chefs as the Japanese market does.

Notes Tetrodotoxin has been found in a broad variety of fish in this family including *Sphoeroides spengleri, Sphoeroides testudineus* (the genus is also spelled *Sphaeroides* and *Spheroides*), the ocean sunfish *Mola mola, Diodon histrix, Lagocephalus leavigatus* and *Chilomycterus schoepfli*. Captain Cook was poisoned from eating the eggs and liver of a *Tetrodon* and survived to provide a detailed account of the symptoms. Peculiarly the same toxin has been found in starfishes, crabs, several marine worms, the venom of the blue-ringed octopus, the skin of salamanders and skin secretions of poison-arrow frogs of the genus *Atelopus*. This broad taxonomic distribution of the toxin has lead to the hypothesis that the toxin might arise from an extrinsic progenitor through the food chain or in symbiotic organisms. Several bacteria commonly found as symbionts have been found to produce tetrodotoxin and several of the normal carriers of the toxin do not produce the toxin when raised in captivity or rendered free of bacteria.

Scombroid poisoning (tuna and mackerel)   p. 110
Scombridae   Fish

Toxic element The toxin is a histamine (sometimes reported as saurine) produced by bacterial action on the flesh of inadequately preserved fish. L-histidine present in the muscles of the fish is converted to histamine by the enzyme histidine decarboxylase, which is produced by certain bacteria, primarily *Proteus, Klebsiella* and *Clostridium*. The bacteria reside in the gills and intestines of the fish and may invade the muscles after death if good hygienic practice and low-temperature storage is not followed. The fish producing the disease are usually various forms of tuna and mackerel, but jacks and others have been known to cause the disease. The disease is associated more frequently with the consumption of raw fish.

Symptoms The onset of symptoms takes place within minutes to at most a few hours after eating the fish. The toxic fish is often noted after the fact to have had a peppery taste. Cutaneous symptoms may include generalized itching, sweating, tingling, burning, swelling and a rash with flushing. Vomiting, diarrhea and abdominal cramps are often accompanied by headache, burning of the throat, numbness and thirst. Palpitations of the heart and low blood pressure due to vasodilation may be present. Symptoms usually subside rapidly and rarely last more than a day. Many outbreaks of histamine poisoning are related to the consumption of raw fish.

Treatment The best treatment is prevention by diligently chilling fish of the tuna and mackerel family promptly after they are caught. The stomach of the patient should be emptied with emetics or lavage if vomiting has not occurred. The disease is self-limiting and mild cases do not require medical intervention. Therapy with H_1 histamine antagonists such as diphenhydramine and chorpheniramine usually results in rapid reduction of symptoms. H_2 antagonists such as cimetidine have also been used effectively.

Beneficial uses Fish of this family include some of the most succulent and abundant commercially harvested species in the world. They should not be avoided, but rather skeptically examined when there is any doubt about handling after capture.

Notes The cero mackerel and kingfish can cause both scombroid poisoning and ciguatera, but the differential diagnosis between the two diseases is quite distinctive.

Marine Toad   p. 110
Bufo marinus   Amphibia

Toxic element Ancient Roman women knew of the toxic properties of toads and applied them to unfaithful husbands. All 12 species of the genus examined contain bufagins (chemically bufadienolides) bufotenines bufotoxins and other compounds. This is the only amphibian genus to produce these toxins. The skin of the back of the toad and the secretions of the parotid glands (the kidney shaped protrusions behind each eye) are the only location of these toxins. The skin and warts on the toad's back contain bufotoxin but none of the other toxins and produces a physiological response in the laboratory almost identical to ouabain. The different bufotoxins are molecular combinations of a specific bufagin with a molecule of erylarginine. The pharmacologic action of bufotoxins and bufagins have a digitalis-like action on the heart with some being more potent than ouabain. Two of the most abundant glycosides are artebufogenin and resibufogenin.

Symptoms When mouthed by a dog, very small amounts of the toxin released from the toad are absorbed by the membranes of the mouth and cause profuse salivation, trembling, heartbeat irregularities, high blood pressure, convulsions and death due to ventricular fibrillation. As might be expected, old dogs with heart problems are considerably more susceptible to the toxin than young healthy animals. Effects on humans are vomiting, increased blood pressure, increased pulse rate, increased rate and depth of respiration, severe headache, and paralysis. In one case of a human hospitalized after the consumption of toad soup serum "digoxin" level measured 2.1 nanograms per ml and was accompanied by cardiopulmonary arrest from which the patient recovered after resuscitation and treatment with atropine. Humans have become seriously intoxicated after the toad poison got into scratches and others have died as a result of eating a meal containing an inadequately prepared toad. One youngster was hospitalized for a week with seizures and other neurological problems after mouthing a toad.

Treatment In cases of suspected *bufo* intoxication the animal's mouth should be thoroughly and vigorously rinsed with a garden hose to remove any toxins remaining. If the dog has been mouthing the toad for some period of time or otherwise is likely to have swallowed a considerable amount of saliva vomiting should be induced to remove as much of the toxin as possible from the stomach. The TDx Digoxin II serum assay has been found to accurately report the effective levels of cardiac glycosides in the blood serum of victims. In dogs the intravenous injection of 5 mg/kg of propranolol hydrochloride brings about rapid recovery. In cases of severe poisoning the dose may need to be repeated after 20 minutes.

Beneficial uses Due to the similarity of action of bufotoxin to ouabain and digitalis it has been recommended in the treatment of heart ailments but does not seem to offer any advantages over more accepted therapies. Dried toad poison has been used effectively in Oriental folk medicine for several thousand years. It is used as a cardiac stimulant diuretic and to treat sinus problems canker sores local inflammations toothache and bleeding from the gums. Toad venom is still used in Chinese pharmacies were it is known as *Ch'an Su*. Several groups of native people have regularly eaten these toads after carefully skinning them and removing the parotid glands.

Notes One of the common names in Hispanic countries is *sapo cururu*. The dried parotid gland of the toad contains bufotenine (Dimethyl-5-hydroxytryptanine) and is smoked to produce psychedelic effects. A brief fad in California involved licking the secretions from the backs of toads for the hallucinogenic effects of the toxins. Surprisingly, no deaths have been reported as a result of this strange adventure. Toad venom is one of the reputed ingredients in the concoction used to produce zombies by practitioners of voodoo in Haiti. The parotid glands may contain over 50 mg of epinephrine. When compared to the adrenals of man with about 8 mg of epinephrine the toad has about 1000 times the amount of epinephrine. A mixture of ouabain and epinephrine acts synergistically and produces symptoms identical to that of toad poisoning in experimental animals. The parotid glands contain cardiac glycosides in a concentration similar to those of the plants oleander and thevetia. The concentration of a digitalis-like substance in the blood serum is equivalent to more than 1000 times the concentration of digitalis that produces toxicity in man. The process of restoring the venom in the parotid glands is very slow. If the contents of the glands are expressed by manual squeezing it takes about 11 weeks to restore 2/3 of the amount of venom. An old French proverb says "this is a wicked beast; if attacked it defends itself"; this is certainly confirmed.

Additional toxins reported from this toad include telocinobufagin marinobufagin (also called bufagin) marinobufotoxin jamaicobufagin resibufagin argentinogenin bufalin hellebrigenol hellebrigenin (also called bufotalidin) and gamabufotalin (also called desacety-cinobufotalin).

Cuban tree frog
Osteopilus septentrionalis

p. 111
Amphibia

Toxic element The skin secretions are irritating.

Symptoms Dogs work their jaws and tongue while salivating heavily often working up a froth of saliva on the lips. A woman experienced severe burning pain in the eye for several hours after handling a tree frog molested by her cat then inadvertently rubbing her eye. The skin secretions can cause a rash on thin skin such as the inside of the forearm.

Treatment Washing out the inside of the dog's mouth with a garden hose seems to eliminate much of the toxin and relieve the symptoms. Careful washing of all parts of the body contacting the frog will prevent symptoms.

Notes When attacked or mauled by cats or dogs this frog gives a loud unmistakable shriek of distress. Many indigenous people in South and Central America use the skin secretions from the small very colorful tree frogs in the family Dendrobatidae as a basis for an arrow poison.

Copperhead
Agkistrodon contortrix

p. 112
Reptilia

Toxic element The venom has a LD_{50} of 10.9 to 27 mg/kg IV in a mouse which makes it about 1/7 as potent as rattlesnake venom. It has a very high protein-digesting activity.

Symptoms The bite is painful but produces less swelling and tissue injury than that of rattlesnakes. The bite and the injection of venom are separate and independent acts under the control of the snake. In about 20% of the confirmed bites by a large poisonous snake no venom is injected and the only symptoms are those of a mechanical puncture wound. The bite with venom injection usually produces an intense pain. With injection of very large amounts of venom the pain may be minimal or absent due to the numbing action of the venom. Redness and swelling coupled with the presence of fang marks are usually definitive of snakebite but confirmed identification of the culprit snake is highly desirable. Tingling numbness and skin discoloration due to subcutaneous hemorrhage in the area of the bite are common. Spontaneous bleeding from the wound may be present. Muscle twitching or paralysis may prevent walking. Internal pain, low blood pressure and weak pulse are typical. The entire limb may show extensive swelling and a black discoloration. Local tissue death may penetrate to the bone or cause the loss of digits. Anemia may be present several days after the bite due to the hemolytic effects of the venom. A medical summary of the bite symptom sequence is pain, edema, alteration of blood characters and necrosis.

Treatment The best treatment is vigilance when walking in woods which are suitable snake habitat. Do not step over a log without looking on the other side and generally pay attention to where you are placing your feet. If bitten the best first aid is use of an automobile to transport the patient to a major hospital. If the bite occurs in a remote area more than 1 hour from medical assistance prompt cut and suction may remove some of the venom. A shallow cut 1/4 inch long can be made parallel to the extremity through each fang mark followed by immediate suction. There are numerous snakebite kits commercially available for this use and it is wise to have one available when traveling in remote country. The polyvalent antivenoms are quite effective at neutralizing the hemorrhagic effects of all members of this family in North America but they are seldom needed as a life-saving measure for bites of this snake except in children or the elderly.

Prompt administration of antivenom greatly alleviates the development of symptoms. (See appendix on antivenom.) Antivenom is most effective when given intravenously within less than 4 hours of the snakebite. As time passes the antivenom is less effective but it has been useful in some cases as long as 24 hours after the bite. Antivenom is a horse-serum product which may produce an adverse reaction in some patients. Intradermal testing for reaction to horse serum should be conducted prior to administration of antivenom and it is preferred that the administration take place in a setting in which epinephrine and resuscitation equipment are available if needed. The treatment of the less severe bites of this species without antivenom has been successful and avoids the risk of serum sickness or anaphylaxis. This treatment includes intravenous fluids, antibiotics (cephelosporins), tetanus, toxoid immobilization, and elevation of the injured part. In patients with coagulation difficulties blood-component replacement therapy may be needed. In some of the older literature the use of ice to chill the bite site and its surroundings to prevent the spread of venom has been recommended. This technique is now generally believed to produce little benefit and in many cases it leads to more extreme tissue damage due to concentration of venom.

Beneficial uses Protac is a commercially marketed single-chain glycoprotein isolated from the venom of copperheads used in blood coagulation research and sensitive assays for protein C.

Cottonmouth
Agkistrodon piscivorus

p. 113
Reptilia

Toxic element The venom has been fractionated in 4 major components 2 of which produce hemorrhages and another with a high proteolytic activity. The venom has the ability to digest blood cells. The venom has a LD_{50} of 4 to 5 mg/kg IV in a mouse which makes it about 40% as potent as rattlesnake venom.

Symptoms The bite and the injection of venom are separate and independent acts under the control of the snake. In about 20% of the confirmed bites by a large poisonous snake no venom is injected and the only symptoms are those of a mechanical puncture wound. The bite with venom injection usually produces an intense pain like a hot needle. With injection of very large amounts of venom the pain may be minimal or absent due to the numbing action of the venom. Redness and swelling coupled with the observation of fang marks are usually definitive of snakebite but confirmed identification of the culprit snake is highly desirable. Tingling numbness and skin discoloration due to subcutaneous hemorrhage in the area of the bite are common. Spontaneous bleeding from the wound may be present. Muscle twitching may be present and paralysis may prevent walking. Internal pain low blood pressure and weak pulse are typical. The entire limb may show extensive swelling and a black discoloration. Local tissue death may penetrate to the bone or cause the loss of digits. Several days after the bite anemia may be present due to the hemolytic effects of the venom. A medical summary of the bite symptom sequence is pain, edema, alteration of blood characters and necrosis. Due to the smaller average size of the snake and relatively less toxic venom the symptoms are usually less severe than those resulting from the bites of rattlesnakes.

Treatment The best treatment is vigilance when walking in the swampy places which are suitable snake habitat. Do not step over a log without looking on the other side and generally pay attention to where you are placing your feet. If bitten the best first aid is use of an automobile to transport the patient to a major hospital. If the bite occurs in a remote area more than 2 hours from medical assistance prompt cut and suction may remove some of the venom. A shallow cut 1/4 inch long can be made parallel to the extremity through each fang mark followed by immediate suction. There are numerous snakebite kits commercially available for this use and it is wise to have one available when traveling in remote country. The polyvalent antivenoms are quite effective at neutralizing the hemorrhagic effects of all members of this family in North America.

Prompt administration of antivenom greatly alleviates the development of symptoms. Antivenom is most effective when given intravenously within 4 hours of the snakebite. As time passes it is less effective but it has been useful in some cases as long as 24 hours after the bite. Antivenom is a horse-serum product which may produce an adverse reaction in some patients. Intradermal testing for reaction to horse serum should be conducted prior to administration of antivenom and it is preferred that the administration take place in a setting in which epinephrine and resuscitation equipment are available if needed. The treatment of the less severe bites of this species without antivenom has been successful and avoids the risk of serum sickness or anaphylaxis. This treatment includes intravenous fluids antibiotics (cephelosporins) tetanus toxoid immobilization and elevation of the injured part. In patients with coagulation difficulties blood-component replacement therapy may be needed. In some of the older literature the use of ice to chill the bite site and its surroundings to prevent the spread of venom has been recommended. This technique is now generally believed to produce little benefit and in many cases it leads to more extreme tissue damage due to concentration of venom.

Fer-de-lance p. 114

Bothrops species Reptilia

Toxic element The venom has a lethal dose of 1.4 to 2.5 mg/kg IV which makes it about as potent as rattlesnake venom. Large snakes may inject over 300 mg of venom per bite. The venom is composed of a multitude of enzymes; some are phosphoesterases and others are proteases with thrombinlike effects.

Symptoms Pain, solid swelling and hemorrhage are evident around the bite site. The venom may at first accelerate clotting with large doses inducing death by massive intravascular clotting. Retardation or complete abolition of clotting due to fibrinolysis occurs later and is usually the most evident influence.. Tissue damage and necrosis so severe as to require amputation may develop in untreated cases. Hemorrhages of the mucous membranes of the nose mouth and gastrointestinal tract may be present in severe bites due to pathological changes in the vessels of the circulatory system. Blood pressure and pulse are generally normal. In severe bites the blood pressure may drop a fever may be present and the blood may become uncoagulable. Edema of the brain, massive intravascular hemolysis, necrosis of the kidneys and uremia may be present in severe cases. Human mortality in untreated cases is about 20%. Overall mortality in which most cases are treated promptly with antivenom is about 0.004%.

Treatment The best treatment is vigilance when walking in woods which are suitable snake habitat. Do not step over a log without looking on the other side and generally pay attention to where you are placing your feet. If bitten the best first aid is use of an automobile to transport the patient to a major hospital. If the bite occurs in a remote area more than 2 hours from medical assistance prompt cut and suction may remove some of the venom. A shallow cut 1/4 inch long can be made parallel to the extremity through each fang mark followed by immediate suction. There are numerous snakebite kits commercially available for this use and it is wise to have one available when traveling in remote country.

Prompt administration of antivenom greatly alleviates the development of symptoms and generally eliminates the need for amputation of necrotic extremities. (See appendix on antivenom.) Antivenom is most effective when given intravenously within less than 4 hours of the snakebite. As time passes the antivenom is less effective but it has been useful in some cases as long as 24 hours after the bite. Antivenom is a horse-serum product which may produce an adverse reaction in some patients. Intradermal testing for reaction to horse serum should be conducted prior to administration of antivenom and it is preferred that the administration take place in a setting in which epinephrine and resuscitation equipment are available if needed. The mortality in untreated cases has varied from 8% to 20%. With antivenom treatment the mortality is about 1% and that is for patients who receive the treatment more than 6 hours after the bite. Death is due to cerebral hemorrhage or collapse of the peripheral circulation. Treatment of the less severe bites of this species without antivenom has been successful and avoids the risk of serum sickness or anaphylaxis. This treatment includes intravenous fluids, antibiotics (cephelosporins),tetanus toxoid,immobilization and elevation of the injured part. In patients with coagulation difficulties blood-component replacement therapy may be needed. Renal failure is not uncommon following *Bothrops* envenomation and should be treated first with rehydration and diuretics. If these fail dialysis may be required until recovery has progressed sufficiently for kidney function to resume. In some of the older literature the use of ice to chill the bite site and its surroundings to prevent the spread of venom has been recommended. This technique is now generally believed to produce little benefit and in many cases it leads to more extreme tissue damage due to concentration of venom.

Beneficial uses The venom is highly suitable as a source material for the preparation of phosphoesterases used in biomedical research.

Notes Over 50,000 *Bothrops* were killed for bounty in Martinique from 1960 to 1966.

Diamondback rattlesnake
Crotalus adamanteus

p. 115

Reptilia

Toxic element The venoms of this genus are a soup of heterogeneous physiologically active compounds including enzymes, glycoproteins, polypeptides and others which have evolved to help the snake capture its prey. The venom produces destruction of tissues, blood cells, membranes and blood-clotting mechanisms. Envenomed victims may show a complete absence of fibrinogen in the blood. Hyaluronidase breaks down the tissue matrix allowing further penetration of the venom into tissue spaces. Swelling may impede circulation to the extent of exacerbating the adverse effects of the venom. The venom has an LD_{50} of 1.2 to 4.2 mg/kg IV in a mouse. As the snake strikes it opens its mouth wide and erects its fangs, aiming them at the victim. Thus the wound is more accurately described as a stab than a bite.

Symptoms The bite and the injection of venom are separate and independent acts under the control of the snake. In about 20% of the confirmed bites by a large poisonous snake no venom is injected and the only symptoms are those of a mechanical puncture wound. Swelling is usually evident within a few minutes and may extend to the entire limb within an hour or it may progress slowly over a period of 8 hours. Without treatment the entire limb may show extensive swelling and a black discoloration. The bite with venom injection usually produces an intense pain like a hot needle. With injection of very large amounts of venom the pain may be minimal or absent due to the numbing action of the venom. Redness and swelling coupled with the presence of fang marks are usually definitive of snakebite but confirmed identification of the culprit snake is highly desirable. Tingling numbness and skin discoloration due to subcutaneous hemorrhage in the area of the bite are common. Spontaneous bleeding from the wound may be present. Muscle twitching may begin within 5 minutes of a bite and paralysis may prevent walking within half an hour. Low blood pressure faintness sweating numbness or tingling (sometimes remote from the bite site) internal pain irregularities of blood characteristics and rapid weak pulse are typical. Local tissue death may penetrate to the bone or cause the loss of digits. Anemia may be present several days after the bite due to the hemolytic effects of the venom. A medical summary of the bite symptom sequence is pain, edema, alteration of blood characters and necrosis. Due to the large amount of venom injected rattlesnakes are responsible for about 95% of the snakebite fatalities in the U.S.

Treatment The best treatment is vigilance when walking in woods which are suitable snake habitat. Do not step over a log without looking on the other side and generally pay attention to where you are placing your feet. If bitten the best first aid is use of an automobile to transport the patient to a major hospital. If the bite occurs in a remote area more than an hour from medical assistance prompt cut and suction (within 5 minutes) may remove some of the venom. A shallow cut 1/4 inch long can be made parallel to the extremity through each fang mark followed by immediate suction. There are numerous snakebite kits commercially available for this use and it is wise to have one available when traveling in remote country. With some of the newly developed suction pumps which produce almost one atmosphere of vacuum up to 23% of the venom can be extracted within 3 minutes. More than 30 minutes after the bite the venom will have diffused into the surrounding tissues and cannot be removed by suction. Alternatively the application of firm pressure over the bite and immobilization of the limb with a splint until antivenom can be infused has been advocated. The polyvalent antivenoms are quite effective at neutralizing the adverse effects of the venom of all members of this family in North America and most of the Crotalidae such as fer-de-lance (*Bothrops*) and bushmaster (*Lachesia*) in South and Central America.

Prompt administration of antivenom greatly alleviates the development of symptoms. (See appendix on antivenom.) Antivenom is most effective when given intravenously within less than 4 hours of the snakebite. As time passes the antivenom is less effective but it has been useful in some cases as long as 24 hours after the bite. Antivenom is a horse-serum product which may produce an adverse reaction in some patients. Intradermal testing for reaction to horse serum should be conducted prior to administration of antivenom and it is preferred that the administration take place in a setting in which epinephrine resuscitation equipment and other resources for treatment of anaphylaxis are available if needed. The antivenom is administered intravenously. The amount is titrated against the signs and symptoms of envenomation and varies from 5 to 45 vials. A useful monitoring technique is to measure the circumference of the extremity adjacent to the bite and at a sequence of points closer to the trunk. Injection of venom directly into the bloodstream is a rare event which produces rapid-onset acute symptoms which may be greatly relieved by the administration of up to 20 vials of antivenom by IV push. About 80% of patients receiving antivenom develop serum sickness while those receiving more than 7 vials inevitably develop serum sickness. In less severe bites the treatment of bite victims without antivenom has been successful. This treatment includes intravenous fluids, antibiotics (cephelosporins), tetanus toxoid, immobilization and elevation of the injured part. In patients with coagulation difficulties, blood-component replacement therapy may be needed. In some of the older literature the use of ice to chill the bite site and its surroundings to prevent the spread of venom has been recommended. This technique is now generally believed to produce little benefit and in many cases it leads to more extreme tissue damage due to long-term concentration of venom.

Beneficial uses The venom has use as an experimental tool as it converts fibrinogen to fibrin directly without decreasing other clotting factors.

Notes The generic name is from the Greek word meaning "rattle" in reference to the rattles on the tail of all members of this genus. The Virginia opossum seems astonishingly immune to the effects of the venom of pit vipers. Further research on the source of this immunity could yield considerable improvement of the normal course of treatment for snakebite.

Coral snake   p. 116
Micrurus fulvius   Reptilia

Toxic element The fangs of coral snakes are much smaller than those of the other poisonous snakes in North America and are situated farther back in the mouth. Coral snakes chew pugnaciously on the victim to work the fangs into the skin before envenomation. Envenomation is rare except for instances when the snake is handled or restrained. The effects of the venom are primarily on the nervous system. The venom has a LD_{50} of 0.2 to 0.4 mg/kg IV in the mouse which makes it about 6 times more potent than rattlesnake venom. Experimental milking of coral snakes has produced four or five adult human lethal doses of venom in a single bite. About 80% of the venom is replaced within 2 weeks after the venom glands have been completely emptied.

Symptoms There are few immediate signs of coral snake envenomation. The pain and swelling typical of pit vipers are usually absent and the neurologic symptoms may not appear for as long as 12 hours after the bite. If the skin is abraded by a snake positively identified as a coral snake it is prudent to assume that a bite has occurred and to monitor the patient carefully in a hospital for at least 24 hours. Even severe envenomation may yield only slight swelling around small fang marks which may release small amounts of blood on squeezing. The bitten limb may feel numb but most of the symptoms are systemic. Labored breathing, excess salivation, weakness, giddiness, apprehension, nausea and vomiting are early symptoms. Paralysis of muscles supplied by the cranial nerves, abnormal reflexes, double vision, muscle fasciculations, difficulty in speaking, depression (or euphoria), and respiratory paralysis may be evident as the venom is distributed through the body. Sudden total paralysis may be delayed for more than 12 hours after the bite. The victim usually remains conscious as long as oxygen exchange is maintained but respiratory paralysis is the typical cause of death from coral snake bite.

Treatment Coral snakes should not be handled by hand. If it is necessary to catch and remove one from the vicinity of a house the snake should be picked up with a rake and dropped into a garbage can for transport. Coral snake bite should be treated as a life-threatening event particularly if the snake "hung on" or chewed. Although the bite may seem minor and the symptoms delayed in onset the symptoms can develop rapidly when they do appear. Thus it has been recommended that antivenom produced specifically for treatment of coral snake bites be located and several vials be administered intravenously to all patients who definitely have been bitten. (See appendix on antivenom.)

A supply of the antivenom is maintained at the Center for Disease Control in Atlanta and at regional poison control centers. Antivenom is a horse-serum product which may produce an adverse reaction in some patients. Intradermal testing for reaction to horse serum should be conducted prior to administration of antivenom and it is preferred that the administration take place in a setting in which epinephrine and resuscitation equipment are available if needed. In the case of a severe anaphylactic reaction to horse serum mechanical maintenance of ventilation until paralysis resolves should be considered as an alternative treatment. Maintaining ventilation will save almost all victims even in the absence of antivenom. Recovery has occurred after as many as 10 weeks of artificial ventilation. Endotracheal intubation is suggested if any signs of bulbar paralysis such as double vision or slurred speech are encountered. The intubation minimizes the probability of aspiration pneumonia and may be required to sustain life in the case of respiratory paralysis. The absolute paralysis usually resolves in less than a week but full muscle strength may not return for a month to 6 weeks.

Notes A rhyme to help remember how to distinguish coral snakes from the other brightly colored, ringed nonvenomous snakes is: Red touch yellow, kill a fellow. Red touch black, a friend of Jack.

Pigmy rattlesnake

Sistrurus miliarius

p. 117
Reptilia

Toxic element The venom has a LD_{50} of 2.8 to 12 mg/kg IV in a mouse which makes it about half as potent as rattlesnake venom but with a much lower volume of venom per bite. While the venom is relatively low in the toxic compound which produces death it is quite potent in destructive effects on tissue near the bite site. This snake is generally less aggressive than the other members of its family and often produces bites with no envenomation.

Symptoms The potential for problems from a bite from this little snake should not be underestimated. The pain is immediate with significant swelling following rapidly. The site of the bite may turn black with swelling and discoloration extending through the entire limb.

Treatment The best treatment is vigilance when walking in woods which are suitable snake habitat. Do not attempt to pick up or handle any unidentified snake encountered in the field. Do not step over a log without looking on the other side and generally pay attention to where you are placing your feet. If bitten the best first aid is use of an automobile to transport the patient to a major hospital. If the bite occurs in a remote area more than 1 hour from medical assistance prompt cut and suction may remove some of the venom. A shallow cut 1/4 inch long can be made parallel to the extremity through each fang mark followed by immediate suction. There are numerous snakebite kits commercially available for this use and it is wise to have one available when traveling in remote country. The polyvalent antivenoms are quite effective at neutralizing the hemorrhagic effects of all members of this family in North America. (See appendix on antivenom.) Antivenom is most effective when given intravenously within less than 4 hours of the snakebite. As time passes the antivenom is less effective but it has been useful in some cases as long as 24 hours after the bite.

Prompt administration of antivenom greatly alleviates the development of symptoms. Antivenom is a horse-serum product which may produce an adverse reaction in some patients. Intradermal testing for reaction to horse serum should be conducted prior to administration of antivenom and it is preferred that the administration take place in a setting in which epinephrine and resuscitation equipment are available if needed. In many of the less severe bites the treatment of bite victims without antivenom has been successful. This treatment includes intravenous fluids antibiotics (cephelosporins) tetanus toxoid immobilization and elevation of the injured part. In patients with coagulation difficulties blood-component replacement therapy may be needed. In some of the older literature the use of ice to chill the bite site and its surroundings to prevent the spread of venom has been recommended. This technique is now generally believed to produce little benefit and in many cases it leads to more extreme tissue damage due to concentration of venom. Fatalities are rarely reported as a result of bites from this snake.

Appendix 1 Antivenom

Theoretical basis for treatment

The concept of immunization against animal toxins by treatment with gradually increasing doses of venom was experimentally demonstrated in the late nineteenth century and has changed little since then. Periodic injection of venom induces the host animal to produce antibodies in its blood serum. Blood serum with antibodies from the immune animal is injected into the host animal or human. Therapy for venomous bites is very different from that for infectious agents. While an infection increases gradually and the causative organism produces increasing amounts of toxin, treatment is intended to reduce the population of the infectious organism and seldom has any direct effect on the toxins. In the case of snakebite the entire dose of toxin is injected at the bite and the antivenom is directed specifically at neutralizing the toxin. It should be remembered that antivenoms only neutralize toxins and do not cure lesions. Prompt treatment with an adequate dose of antivenom is essential in preventing further development of symptoms.

Production

Venom is milked from snakes by mechanical pressure on the venom gland or electrical stimulation of the muscles surrounding the gland. The glands recharge in 3 to 4 weeks allowing repeated milking of snakes held in captivity. There is some geographic variation in the venom produced as well as variations between individual snakes making it advantageous to use venom pooled from different snakes as the stimulus for antibodies. Venom may be used fresh but now it is most commonly dehydrated by the venom supplier and later reconstituted for use by the laboratory which produces antivenom. To reduce the inventory of antivenom needing to be maintained at hospitals or treatment centers, polyvalent antivenom is commonly produced. The antibody-producing animal is injected with the venom from several types of snakes which have similar characteristics to produce polyvalent antivenom.

The standard host animal for the production of antibodies is the horse because it is easy to maintain and yields large amounts of serum. Donkeys, mules, cows, and oxen have been used but are not generally in commercial production. Goats, sheep and rabbits also produce satisfactory antibodies but are generally used only in experimental procedures. Various methods have been devised to reduce the toxicity but maintain the antigenic properties of venom but in every case a non–life threatening amount of blood is withdrawn from the host animal. The whole blood from the immunized animal is separated from the serum. The blood serum is refined, concentrated ,freeze-dried, and packaged as an antivenom which is reconstituted just prior to its injection into a victim of snakebite.

The decision to use antivenom should be made after considering the following factors.

1. The identification of the snake or other venomous animal.
   The bite may not be from a dangerous animal.

2. Clinical features demonstrated by the victim of the bite.
   The animal may have bitten without injecting a significant amount of venom.

3. Time elapsed since the bite.
   The effectiveness of the antivenom in reducing symptoms declines with time after the bite.

4. The sensitivity of the victim to horse serum.
   The cost/benefit ratio may suggest the withholding of antivenom in the case of a mild bite to a patient with great allergic sensitivity to horse serum.

Administration

Antivenom should be administered in a hospital setting if at all possible due to possible adverse reaction of the patient to horse serum. The antivenom is usually administered as an addition to an IV drip because adverse reactions are reduced and drugs for symptomatic treatment can be included. Under emergency circumstances antivenom may be injected slowly directly into a vein. Intramuscular or subcutaneous injection at the bite site is not recommended as it may take up to 4 hours to diffuse into the circulatory system. The following factors should be considered when administering antivenom.

1. The antivenom should be the appropriate one for the species of snake which inflicted the bite.

2. Early administration of antivenom is essential in preventing and reducing adverse effects of the bite.

3. A sufficient dose of antivenom is essential. If sufficient antibodies are injected to neutralize only half the venom the remaining venom will remain active and continue to produce adverse effects.

4. The entire dose at once will have the most beneficial effect in preventing further development of symptoms produced by the venom.

5. The injection route should be chosen to promptly introduce the antivenom into the general circulation.

Appendix 2 Toxins and Other Chemicals of Interest

Abrin The primary toxin in *Abrus precatorius* is a series of glycoproteins each with a molecular weight of about 65000.

Abrine One of the toxins of *Abrus precatorius* which chemically is N-methyl-L-tryptophan with the molecular formula $C_{12}H_{14}N_2O_2$.

Acetyalloyohimbine An alkaloid from the root bark of *Rauvolfia* with the formula $C_{28}H_{28}N_2O_4$.

Allelopathic A term used by biologists to describe a chemical which inhibits plant growth. Certain plants produce allelopathic chemicals to insure that they are not crowded by competitors.

Allocryptopine An alkaloid from *Argemone* with the formula $C_{21}H_{23}O_5N$ and analgesic, soporific and antiarrhythmic action. It occurs in two allotropic forms: alpha-fagarine or fagarine I and fagarine II. Fagarine I slows the heart and increases the amount of current necessary to provoke fibrillation. It has been used with good results in patients suffering from auricular fibrillation and flutter. Fagarine II lowers the blood pressure and is a respiratory stimulant and muscle relaxant.

Alpha-solanine The primary toxic compound in the genus *Solanum* with the formula $C_{45}H_{73}O_{15}N$.

Anabasine One of the primary toxic alkaloids in *Nicotiana*. It has the formula $C_{10}H_{14}N_2$ and has an LD100 subcutaneously in guinea pigs of 22 mg/kg. It is also called neonicotine and affects the nervous system producing confusion, salivation, disturbed vision, dizziness, diarrhea, and convulsions.

Anacardic acid One of the primary irritants making up 90% of the shell oil of *Anacardium*. It has the formula $C_{22}H_{32}O_3$.

Anacrotine A pyrolizzidine alkaloid from the seeds of *Crotalaria verrucosa*. It has an antitumor action.

Andirine An alkaloid from *Andira* also called angelin.

Atropine The primary pharmacologically active drug in *Datura* with the formula $C_{17}H_{23}O_3N$. It is also called hyoscyamine and daturine. It causes blurred vision suppressed salivation vasodilation and agitation.

Asclepiadin A glucoside from *Asclpias curassavica* with cardioactive and emetic action.

Azaridine An alkaloid from *Melia*.

Azadirachtin A triterpene with the formula $C_{35}H_{44}O_{16}$ from *Melia azedarach*. It is toxic to most insects and is used as a deterrent against several pests.

Berberine A bitter yellow alkaloid with the chemical formula $C_{20}H_{19}O_5N$ from *Argemone* and from *Andira*. It probably is the same chemical as argemonine. Its physiological effects are to lower the blood pressure dilate the peripheral blood vessels drop the body temperature and depress the respiratory system. Medically berberine has been found effective in treating *Trypanosoma leishmania* oriental sores (varieties of which are also known as tropical sores or Leishmania) and trachoma. It is a provocative agent which liberates the malaria parasite into the peripheral circulation making it more susceptible to diagnosis and treatment. It also has anticonvulsant antipyretic and sedative properties.

Brevitoxin The nonprotein toxin produced by *Gymnodinium breve* with the formula $C_{50}H_{70}O_{14}$. The toxin is actually a series of compounds with a common structure of contiguous fused ether rings.

Bufagin One of the primary toxins found in the parotid glands of the toad *Bufo marinus*. It has the formula $C_{28}H_{36}O_6$. Doses of more than 0.4 mg/kg are fatal in cats. Orally it has a bitter taste and produces a numbness of the tongue.

Bufalin A cardiotonic steroid refined from Bufo venom. It is listed as a U.S.P. cardiotonic and 2 U.S. patents have been issued on its preparation for pharmaceutical use.

Bufogenin A toxin with the formula $C_{24}H_{34}O_5$ isolated from Bufo venom.

Bufotoxin One of the primary toxins found in the parotid glands of the toad *Bufo marinus*. It has the formula $C_{38}H_{58}O_{10}N_4$. Doses of more than 0.29 mg/kg are fatal in cats. Orally it produces a numbness of the tongue.

Cajeput A volatile oil distilled from *Melaleuca* leaves and stems. It contains 50% to 60% cineol (also called eucalyptol) along with pinene and terpinol. This oil has an allergic cross-reaction with eucalyptus oil which contains up to 80% cineol. Commercial versions of it often have a green tinge from the copper salts of the crude copper stills used to produce the oil.

Calactin One of the cardiotoxic glycosides in *Calotropis procera* with the for mula $C_{29}H_{38}O_9$.

Calotropin One of the cardiotoxic glycosides in *Calotropis procera* and sever al other genera of the Asclepidaceae. The chemical formula is $C_{29}H_{40}O_9$. The LD_{50} in cats is 0.12 mg/kg. It also is used as an arrow poison in Africa.

Calotoxin One of the cardiotoxic compounds in *Calotropis procera* with the formula reported as $C_{20}H_{40}O_{10}$ or $C_{32}H_{42}O_8NS$.

Calotropogenin One of the cardiotoxic compounds in *Calotropis procera* with the formula $C_{23}H_{32}O_6$.

Canela One of the alkaloids from *Canella*.

Cannelal An alkaloid from Cannella with antifeedant antiseptic and fungi cidal properties. It is used as a spice and as an aromatic additive to smoking tobacco.

Cardol An irritant resin-like compound with the formula $C_{21}H_{30}O_2$ found in *Schinus, Anacardium* and several other members of the Anacardiaceae.

Chaconine One of the toxic steroidal glycoalkaloids in the plant family *Solanaeceae*. It is very similar to solanine.

Chlorogenic acid Allergen in *Cestrum nocturnum* with antifeedant antioxi
dant and cancer-preventive properties. Chemically it is 3-(34-dihydrox-
ycinnamoyl) quinic acid with the formula $C_{16}H_{18}O_9$.

Chrysarobin The toxin from *Cassia occidentalis*. This yellow-brown powder
also is derived from the wood of *Andira araroba* and is used medicinally as
a fungicide to treat skin diseases such as ringworm. It is also used as an
emetic and irritant and to treat psoriasis. It is marketed as an extract from
Andira which contains a series of anthraquinone derivatives. It is also synthesized as a reduction product of chrysophic acid with the formula
$C_{15}H_{12}O_3$.

Ciguatoxin A lipid-soluble heat-stable toxin with the tentative formula
$C_{53}H_{77}NO_{24}$. It is produced by a dinoflagellate and is concentrated as it
passes up the food chain and accumulates in carnivorous marine fishes.

Cissampeline The primary active alkaloid in the root of *Cissampelos pareira*
with the formula $C_{19}H_2NO_3$.

Cissampareine One of the toxic antitumor alkaloids in the root of
Cissampelos pareira. It has the formula $C_{37}H_{38}N_2O_6$.

Colchicine An alkaloid amine acetyltrimethylcolchicinic acid with the for
mula $C_{22}H_{25}O_6N$ found in *Gloriosa superba*. It is able to withstand lengthy
storage, drying and boiling without detoxification. It is cytotoxic and stops
cell division at mitosis, making it useful in the treatment of leukemia. It
has been used effectively for many years in the treatment of gout but its
mechanism of action is unknown.

Coptisine An alkaloid from *Argemone* with the formula $C_{19}H_{15}O_5N$.

Coronaridine A cytotoxic alkaloid from *Ervatamia* which also has diuretic
analgesic antifertility and central nervous system effects.

Crotalaburnine Toxin in seeds of *Crotalaria verucosa* which is used to reduce
swelling and inflammation and as an anesthetic.

Crotaverrine One of the alkaloids in *Crotalaria verrucosaas*.

Croton oil The purgative skin-irritating and cocarcinogenic oil present in
the seeds and sometimes other parts of the genus *Croton*. The oil contains
several different compounds which account for its properties. One compound is nonaromatic and has the formula $C_{36}H_{56}O_8$. Another compound consists of 3 components two of which are esters of the diterpenoid
alcohol with the formula $C_{20}H_{28}O_6$.

Crotonosine An alkaloid in *Croton discolor.*

Curcin The toxic protein from the seeds of plants of the genus *Jatropha*. A
synonym for jatrophin.

Cycasin The primary toxin (chemically methylazoxymethoxyglucoside)
from *Zamia* with the formula $C_8H_{16}N_2O_7$. It is hydrolyzed in the digestive tract to produce the physiologically active methylazoxymethanol
which is carcinogenic but has been found to have antiaging effects in lab
animals.

Deserpidine A pharmacologically active alkaloid named 11-demethoxy reserpine from *Rauwolfia* with the formula $C_{32}H_{38}N_2O_8$.

Dexamethasone A synthetic adrenocortical steroid anti-inflammatory drug sold under many trade names.

Diterpene esters A series of complex chemicals synthesized by the Euphorbiaceae.

Dihydrosanguinarine This nontoxic alkaloid with the formula $C_{20}H_{15}NO_4$ makes up 85% of the alkaloid content of *Argemone* seed oil.

Dioscorine One of the toxic alkaloids from *Dioscorea* with the formula $C_{13}H_{19}NO_3$. Its actions are anesthetic antidiuretic paralytic convulsant and mydriatic.

Diosgenin The aglycone of the saponin disocin which can be used as a source material for the production of the reproductive hormone progesterone. It has the formula $C_{27}H_{42}O_3$ and is antifatigue anti-inflammatory antistress and estrogenic.

Diphenhydramine hydrochloride An antihistamine drug also used to pre vent motion sickness. It is sold under many trade names.

Discolorine An alkaloid in *Croton discolor*.

Dregamine An alkaloid from *Ervatamia* with the chemical formula $C_{21}H_{26}N_2O$. It is a respiratory stimulant, local anesthetic and produces convulsions at high doses.

Ervaticine An alkaloid from *Ervatamia* with the formula $C_{17}H19NO_2$.

Erysotrine A neuromuscular blocking agent from the leaves of *Erythrina* with the formula $C_{19}H_{23}NO_3$.

Erythrinine A toxic alkaloid in *Erythrina*.

Erythroidine A neuromuscular and ganglionic blocking alkaloid with an action similar to curare. It has potential in medicine as a skeletal muscle relaxant. It is found in the seeds of *Erythrina* and has the formula $C_{16}H_{19}NO_3$. The LD_{50} IV in rabbits is 8.6 mg/kg.

Euphol The toxic triterpene resin found in the latex of *Euphorbia* with the formula $C_{30}H_{48}O$.

Euphorbol The toxin in Pedilanthus and other *Euphorbia* which has been used as an abortifacient.

Euphorbon One of the toxins in the latex of *Euphorbia* with the formula $C_{29}H_{36}O_7$. It is also called euphorbin and euphorbiosteroid.

Gelsemicine The toxic alkaloid from *Gelsemium* that is a depressant which lowers the blood pressure and may produce respiratory paralysis.

Gelsemine A toxic alkaloid with the formula $C_{20}H_{26}N_2O$ from the roots of *Gelsemium*.

Gloriosine A toxic alkaloid from roots of *Gloriosa superba* with the formula $C_{33}H_{38}N_2O_9$.

Gossypol The toxic polyphenolic compound with the formula $C_{30}H_{30}O_8$ from the seeds and root bark of *Gossypium*. The LD_{50} is 20 mg/kg in rats

I.P. The proper chemical name for the compound is 11'66'77'-hexahy-droxy-55' diisopropyl-33' dimethyl (22'-binapthalene)-88'-dicarboxyalde-hyde.

Hayatine The predominant alkaloid in *Cissampelos* with muscle-relaxant antitumor respiratory-stimulant and blood-pressure-lowering properties.

Hecogenin A steroidal sapogenin from many *Agave* species. It is used as a source material for manufacture of steroidal hormones.

Hecubine An alkaloid found in the leaves and flowers of *Ervatamia* with the formula $C_{20}H_{26}N_2O$.

Heliotrine An alkaloid which causes veno-occlusive disease. It lowers the blood pressure and has been used experimentally to treat tumors and leukemia.

Hippomanin A water-soluble compound extracted from the leaves of *Hippomane mancinella* which forms pale-yellow crystals and has the for-mula $C_{27}H_{22}O_{18}$.

Hurain A protease found in *Hura crepitans* seeds. It is inhibited by trypsin.

Huratoxin A piscicidal and cocarcinogenic compound from *Hura crepitans* with the formula $C_{34}H_{48}O_8$.

Hydrocyanic acid An extremely toxic compound which interferes with oxy gen transport by the blood. It has the formula HCN with an LD_{50} of 3.7 mg/kg and is found in numerous plants. The specific antidote is sodium thiosulfate.

Hyderabadine An alkaloid from *Ervatamia* leaves with the formula $C_{21}H_{28}N_2O_2$.

16-Hydroxyphorbol The toxic component of *Aleurites* (tung) seeds.

Hyoscine One of the alkaloids in *Datura* also called scopolamine.

Hyoscyamine The primary alkaloid in *Datura stramonium*. Also called atropine.

Hypoglycin A The cyclopropanoid amino acid toxin found in the seeds and unripe fruits of *Blighia sapida*. It is the chemical L-(methylenecyclopropyl) alanine with the formula $C_7H_{11}NO_2$. It has also been reported as alpha-amino-beta-(2-methylenecyclopropyl) proprionic acid.

Hypoglycin B The toxic dipeptide derivative ($C_{12}H_{18}N_2O_5$) of hypoglycin A. It contributes to the toxicity of the seeds of *Blighia sapida*.

Indican A toxic glucoside with the formula $C_{14}H_{17}NO_6$ from *Indigofera*. With hydrolysis it yields glucose and indican which upon oxidation yields the indigo of commerce.

Indicaxanthin One of the suspect toxic alkaloids in *Mirabilis jalapa* with the formula $C_{14}H_{16}N_2O_6$.

Indigotin A toxic glucoside from *Indigofera* with the formula $C_{16}H_{10}O_2N_2$. This is the compound known as indigo in commerce and history.

Indomethacin An anti-inflammatory antipyretic analgesic used as one of the treatments for *Dieffenbachia* intoxication.

Ingenol esters A group of similar irritating and cocarcinogenic chemical compounds derived from the diterpene ingenane.

Integerrimine A toxic alkloid from *Crotalaria incana* with the formula $C_{18}H_{25}O_5N$.

Jatrophin A synonym for curcin.

Jatrophine A bitter toxic alkaloid from the roots and bark of *Jatropha gossyp ifolia* with the formula $C_{14}H_{20}O_6N$ and action against malaria similar to that of quinine.

Jatrophone A diterpene from *Jatropha* which is active against leukemia in mice and human nasal passage carcinoma. It has the chemical formula $C_{20}H_{24}O_3$.

Lachnanthocarpone One of the toxic constituents of *Lacnanthes* with the formula $C_{19}H_{12}O_3$.

Lantadene A toxic triterpene from *Lantana* with the formula $C_{35}H_{52}O_5$. The compound is found in two forms: Lantadene A (also called lantanine or lanthanine) and Lantadene B which differ only in the structure of the molecule.

Lectin A class of toxic plant proteins which bind specifically to carbohydrates. The toxic action is a result of the inhibition of ribosomal protein synthesis.

Leonotin One of the toxic diterpenes in *Leonotis* with the formula $C_{20}H_{30}O_6$.

Lidocaine hydrochloride A local anesthetic drug sold under the trade names of Xylocaine hydrochloride and Anestacon.

Linearisine An alkaloid in *Croton discolor*.

Louisfieserone A prenylflavone from *Indigofera* with the formula $C_{22}H_{24}O_5$.

Lupeol A terpinoid with the formula $C_{30}H_{50}O$ found in the latex of *Ervatamia* and several other plants.

Lycorine A biologically active alkaloid from *Zephyranthes* with the formula $C_{16}H_{17}O_4N$ which is analgesic, antiviral, antimitotic, and reduces fevers.

Margosine A bitter alkaloid from the bark of *Melia*.

Mammein The principal insecticidal constituent in the seeds and fruits of *Mammea americana*. It has the chemical formula $C_{22}H_{28}O_5$ and structurally is 4-n-propyl-57-dihydroxy-6-isopentenyl-8-isovalerylcoumarin.

Mangrovin A bitter component of the bark of *Melia azedarach* used to control intestinal worms.

Mehranine An alkaloid from the leaves of *Ervatamia* with the formula $C_{20}H_{26}N_2O$.

Melittin A hemolytic polypeptide with the formula $C_{131}H_{229}N_{39}O_{31}$ making up 50% of the venom of honey bees. The other 50% is phospholipase A. It has been used as an antirheumatic drug.

Meperidine hydrochloride A narcotic analgesic drug sold under the trade name of Demerol hydrochloride and others.

Meteloidine One of the pharmaceutically active alkaloids in *Datura* with the formula $C_{13}H_{21}O_4N$.

Methylcrotonosine An alkaloid in *Croton discolor*.

Milliamine The primary toxic alkaloid of *Euphorbia milii* with the formula $C_{43}H_{47}N_3O_9$.

Mimosine A toxic amino acid found only in the genera *Leucaena* and *Mimosa*. Chemically it is beta-N(3-hydroxy-4-pyridone)-alpha-aminopropionic acid. It may be present in concentrations of up to 5% dry weight in the young leaves.

Monocrotaline The major toxic pyrrolizidine alkaloid in *Crotalaria retusa* and several other *Crotalaria* with the formula $C_{16}H_{23}O_6N$. It is the monocrotalic acid ester of retronecine and is active as a male insect sterilant and an antitumor agent against adenocarcinoma. It is depressant to the heart and toxic to the liver. It is very stable and has been isolated from the livers of animals poisoned by *Crotalaria*.

Mucanain The itch-producing chemical associated with the hairs of *Stizolobium pruriens*. It is a heat-sensitive protein with a molecular weight of 40,000.

Nepetaefolinol One of the toxic diterpenes in *Leonotis* with the formula $C_{20}H_{28}O_6$.

Nepetaefuran One of the toxic phenols in *Leonotis* with the formula $C_{22}H_{28}O_7$.

Nepetaefuranol One of the toxic phenols in *Leonotis* with the formula $C_{20}H_{30}O_8$.

Neriifolin A cardiac glycoside from *Thevetia* with the formula $C_{33}H_{61}O_{30}$ also reported as $C_{30}H_{46}O_8$.

Nerinine A toxic compound found in *Zephyranthes*.

Nicotine A toxic alkaloid with the formula $C_{10}H_{14}N_2$ from *Nicotiana*.

Nor-hyoscyamine One of the active alkaloids in *Datura* with the formula $C_{17}H_{21}O_3N$.

Oleandrin One of the major cardiac glycosides of *Nerium* with the formula $C_{32}H_{48}O_9$. It is also known as oleandroside.

Peruvoside A therapeutic cardioactive glycoside from *Thevetia* with the formula $C_{30}H_{44}O_{10}$ and an action similar to digitalis. The LD_{50} in cats is 0.15 mg/kg.

Phellandrene The primary toxic terpene found as a nitrate in the fruits of *Schinus terebinthifolius* with the formula $C_{10}H_{16}N_2O_3$. It is used commercially in fragrances but causes vomiting and diarrhea if taken internally.

Phenobarbital sodium An anticonvulsant hypnotic sedative drug sold under several trade names. Regular use may lead to habituation or addiction.

Phylanthin A crystalline bitter compound with the formula $C_{30}H_{37}O_8$ from the bark of *Phyllanthus niruri*.

Phorbol esters Also called tigliane esters. Most of this group of compounds is irritating and cocarcinogenic.

Protopine An alkaloid from *Argemone* with the formula $C_{20}H_{19}O_5N$ which is also called fumarine or macleyine. It produces disturbance of the heartbeat and has sedative muscle-relaxant and analgesic effects. It induces strong uterine contractions and has been used to induce abortions.

Pterygospermin An antibiotic substance from *Moringa oleifolia* with the for mula $C_{22}H_{18}O_2N_2S_2$. The pharmacological antibiotic effect is dependent on the freed benzylisothiocyanate. The LD_{50} in mice is 400 mg/kg orally.

Quercetin Found widely in rinds of fruit and barks of plants. It is used ther apeutically to reduce capillary fragility. It has antihistamine antianaphylactic and antitumor effects. The oral LD_{50} in mice is 160 mg/kg.

Raunitidine A physiologically active alkaloid from the leaves of *Rauvolfia nitida*. It is an epimer of reserpine.

Raunescine One of the physiologically active alkaloids from *Rauvolfia* with the formula $C_{31}H_{36}N_2O_8$.

Rauwolsine The major alkaloid present in the plant *Rauvolfia nitida* with the formula $C_{21}H_{26}N_2O_3$.

Retronecine The most common base portion of pyrrolizidine alkaloids from *Crotalaria*. The formula is $C_8H_{13}NO_2$ and it has an LD_{50} in mice of 634 mg/kg.

Retrorsine One of the toxic alkaloids of *Crotalaria retusa* with the formula $C_{18}H_{25}O_6N$.

Ricinin An alkaloid in *Ricinus*.

Ricinoleic acid The cathartic-producing component of castor oil derived from the seeds of *Ricinus communis*. It has the formula $C_{17}H_{32}(OH)COOH$. It makes up 90% of the triglycerides in castor oil and has been used in contraceptive jellies.

Rivianin The pigment in the berries of *Rivina* with the formula $C_{24}H_{26}N_2O_{16}S$.

Ruvoside A toxin from *Thevetia* with the reported chemical formula $C_{30}H_{46}O_{10}$.

Sanguinarine An alkaloid with the formula $C_{20}H_{14}NO_4$ making up 5% of the alkaloid content but providing most of the toxicity of the seed oil of *Argemone mexicana*. It is also anesthetic antiseptic and fungicidal.

Schinol An irritant triterpene with the formula $C_{30}H_{50}O_3$ from the fruit of *Schinus terebithifolius*.

Scolopetin One of the alkaloids in *Datura stramonium* also known as chrysatropic or gelseminic acid with the formula $C_{10}H_8O_4$.

Scopolamine One of the pharmaceutically active drugs in *Datura* with the formula $C_{17}H_{21}O_4N$. It is also called hyoscine.

Sesbanimide A The primary toxin from *Daubentonia* with the formula $C_{15}H_{21}NO_7$.

Siamine A toxic alkaloid in *Cassia siamea* with the formula $C_{11}H_{11}NO_3$.

Solandrine The toxic alkaloid in *Solandra*. It may be identical to solasonine.

Solanidine A toxic alkaloid from the genus *Solanum* with the formula $C_{27}H_{43}NO_2$. It is a product resulting from the hydrolysis of solanine.

Solanine The toxic glycoalkaloid in Solanaceae with the formula $C_{45}H_{73}O_{15}N$. It reduces the symptoms of parkinsonism and epilepsy but produces defects in fetal development.

Solasodine A steroidal alkaloid from Solanum with the formula $C_{27}H_{43}NO_2$. It is a hydrolysis product of solasonine.

Solasonine A toxic alkaloid with the formula $C_{45}H_{73}O_{16}N$ from several species of *Solanum*.

Solenamine The toxin from the venom of Solenopsis composed of 2-methyl-3-hexadecyl-pyrrolidine and 2-methyl-3-hexadecyl-3-pyrroline.

Stapfinine An alkaloid from the leaves of *Ervatamia* with the formula $C_{19}H_{24}N_2O_2$.

Superbine An alkaloid from *Gloriosa superba* with the formula $C_{52}H_{66}O_{17}N_2$.

Tannin Complex polyphenol mixtures with molecular weights between 500 and 3000. They are often found in the roots and bark of plants.

Tazattine One of the toxic alkaloids from *Melia* with the formula $C_{18}H_{21}NO_5$.

Terebinthone An irritant triterpene with the formula $C_{30}H_{46}O_3$ from the fruit of *Schinus terebithifolius*.

Tetrodotoxin The toxin found in the internal organs of several types of puffer fish. It is a perhydroquinazoline with the formula $C_{11}H_{17}O_8N_3$. The toxin is unstable and decomposes at a pH below 3 or above 7.

Thevetin A toxic compound from *Thevetia* with the formula $C_{29}H_{46}O_{12}$-$2H_2O$. It has been used in Europe and the Soviet Union as a therapeutic heart drug similar in action to digitalis. It has an LD_{50} in cats of 1.1 mg/kg.

Thevetoxin A cardiac glucoside from *Thevetia* with the formula variously reported as $C_{20}H_{30}O_6$ or $C_{30}H_{46}O_{10}$.

Tiglane See phorbol.

Triamcinolone acetonide An anti-inflammatory corticosteroid drug marketed under several names.

Trigonelline One of the alkaloids extracted from *Mirabilis* with the formula $C_7H_7O_2N$.

Tung The seed kernel contains 2 toxins with the formulas $C_9H_{14}O_2$ and $C_{11}H_{16}O_3$.

Ursolic acid A triterpene with the formula $C_{30}H_{48}O_3$ found in *Ervatamia* and *Thevetia*. It is also found in the waxy coating of pears, apples, and prunes. It is active against cancer cells in laboratory experiments.

Urushiol The resinous toxic component of many members of the Anacardiaceae. Chemically it is a 3-n-pentadecylcatechol.

Uscharin One of the cardiotoxic glycosides in *Calotropis procera* with the formula $C_{31}H_{41}O_8NS$.

Voruscharin One of the cardiotoxic glycosides in *Calotropis procera* with the formula $C_{33}H_{47}O_9NS$.

Selected References

Alexander, J.O. 1984. *Arthropods and Human Skin*. Berlin: Springer-Verlag

Allen, O.N. and E.K. Allen. 1981. *The Leguminosae: A Sourcebook of Characteristics, Uses and Nodulation*. Madison, WI: University of Wisconsin Press.

Arena, J. 1974. *Poisoning: Toxicology, Symptoms, Treatment*. Springfield Ill: Charles C. Thomas.

Ayensu, E.S. 1981. *Medicinal Plants of the West Indies*. Algonac MI: Reference Publications.

Behl, B.N. 1966. *Skin Irritant and sensitizing plants found in India*. New Delhi India: Medical College:

Blackwell, W.H. 1990. *Poisonous and Medicinal Plants*. Englewood Cliffs, NJ: Prentice Hall.

Blohm, Henrik. 1962. *Poisonous Plants of Venezuela*. Cambridge MA: Harvard University Press.

Bucherl, W. and E. Buckley (Editors) 1971. *Venomous Animals and Their Venoms*. New York: Academic Press

Correll, D. S. and H. B. Correll. 1982. *Flora of the Bahama Archipelago*. Vaduz, Germany: J. Cramer.

Covacevich, J., P. Davie and J. Pearn. (Editors) 1987. *Toxic Plants and Animals: A Guide for Australia*. Brisbane: Queensland Museum

Driesbach, R.H. 1987. *Handbook of Poisoning: Prevention, Diagnosis and Treatment*. Los Altos, CA: Lange medical publishers.

Duke, J.A. and E.S. Ayensu. 1985. *Medicinal Plants of China*. Algonac MI: Reference Publications.

Duke, J. A. 1985. *CRC Handbook of Medicinal Herbs*. Boca Raton, Fl: CRC Press.

Ellis, M.D. 1975. *Dangerous Plants, Snakes, Arthropods and Marine Life of Texas*. Galveston: U.S. Government Printing Office.

Goddard, J. 1993. *Physicians Guide to Arthropods of Medical Importance*. Boca Raton, Fl: CRC Press.

Halstead, B.W. 1988. *Poisonous and Venomous Marine Animals of the World*. Darwin Press. Princeton NJ

Hardin, J. and J. Arena. 1974. *Human Poisoning from Native and Introduced Plants*. Durham NC: Duke University Press.

Harris, J.B. (Editor) 1978. *Natural Toxins-Animal, Plant and Microbial*. Oxford: Clarendon Press.

Harwood, R.F. 1979. *Entomology in Human and Animal Health*. New York: Macmillan

Howard, R.A. et. al. 1974-1989 (6 vols.) *Flora of the Lesser Antilles*. Jamaica Plain MA: Arnold Arboretum, Harvard University.

Humphreys, D.J. 1988. *Veterinarian Toxicology*. London: Bailliere Tindall.

Keegan, H.L. 1980. *Scorpions of Medical Importance*. Jackson MI. University Press of Mississipi.

Keegan, H.L. and W.V. McFarlane (Editors). 1963 *Venomous and Poisonous Animalx and Noxious Plants in the Pacific Region*. Oxford: Pergamon Press.

Keeler, R. and A. Tu. (Editors). 1983. *Handbook of Natural Toxins*. New York: Marcel Dekker

Kinghorn, D.A. (ed) 1979. *Toxic Plants*. New York: Columbia University Press.

Kingsbury, J.M. 1964. *Poisonous Plants of the U.S. and Canada*. Englewood Cliffs, NJ: Prentice Hall.

Klauber, L.M. 1972. *Rattlesnakes*. Berkeley: University of California Press.

Lampe, K. and R. Fagestrom. 1968. *Plant Toxiciy and Dermatitis*. Baltimore: Williams and Wilkins.

Lampe, K.L. and M.A. McCann. 1985. *AMA handbook of Poisonous and Injurious Plants*. Chicago: American Medical Assoc.

Lee. C.Y. (Editor). 1979. *Snake Venoms*. Berlin: Springer-Verlag

Lewis, W.H. and M.P.H. Elvin-Lewis. 1977. *Medical Botany*. New York: John Wiley.

Little, E.L. and F.H. Wadsworth. 1964. *Common trees of Puerto Rico and the Virgin Islands*. Washington DC: U.S.D.A. Forest Service.

Little, E.L., R.O. Woodbury, and F.H. Wadsworth. 1974. *Trees of Puerto Rico and the Virgin Islands*. Vol. 2. Washington DC: U.S.D.A. Forest Service.

Mitchell, J. and A. Rook. 1979. *Botanical Dermatology: Plants injurious to the skin*. Vancouver: Greengrass

Morton, Julia F. 1977. *Major Medicinal Plants: Botany, Culture, and Uses*. Springfield IL: Charles Thomas.

Morton, Julia F. 1977. *Poisonous and injurious higher plants and fungi*. In: Tedeschi, C. et. al. (Editors) *Forensic Medicine*. Philadelphia: W.B. Saunders.

Morton, Julia F. 1981. *Atlas of Medicinal Plants of Middle America Bahamas to Yucatan*. Springfield IL: Charles Thomas.

Oakes, A. J. and J. O. Butcher. 1962. *Poisonous and Injurious Plants of the U.S. Virgin Islands*. Washington DC: U.S.D.A. Agriculture Research Service. Misc. Pub. 882.

Oliver-Bever, B. 1986. *Medicinal Plants in Tropical West Africa*. New York: Cambridge University Press.

Perkins, K.D. and W.W.Payne. 1978. *Guide to the Poisonous and Irritant Plants of Florida*. Gainesville: Univ of Fl.

Polis, G.A. 1990. *The Biology of Scorpions*. Stanford CA: Stanford University Press

Russell, F.E. and P.R. Saunders (Editors). 1971. *Venomous animals and their venoms*. Academic press: New York.

von Reis, S, and F.J. Lipp Jr. 1982. *New plant Sources for Drugs from the New York Botanical Garden Herbarium.* Cambridge MA: Harvard University Press.

Schultes, R.E. and A. Hoffmann. 1980. *The Botany and Chemistry of Hallucinogens.* Springfield IL: Charles Thomas.

Seaforth, C.E., C.D. Adams and Y. Sylvester. 1983. *A Guide to the Medicinal plants of Trinidad and Tobago.* Port of Spain, Trinidad: Commonwealth Secretariat.

Sonenshine, D.E. 1991. *Biology of Ticks.* London: Oxford University Press.

Tang, W. and G. Eisenbrand. 1992. *Chinese Drugs of Plant Origin.* London: Springer Verlag.

Tu, A.T. (Editor). 1984. *Handbook of Natural Toxins.* New York: Marcel Dekker.

Watt, J. and M. Breyer-Brandwijk. 1962. *Medicinal and Poisonous Plants of Southern and Eastern Africa.* 2nd edition. Edinburgh and London: E&S Livingstone.

Index